MW00582913

NCHRP REPORT 489

Design of Highway Bridges for Extreme Events

MICHEL GHOSN
FRED MOSES

AND

JIAN WANG
City College
City University of New York
New York, NY

SUBJECT AREAS
Bridges, Other Structures, and Hydraulics and Hydrology

Research Sponsored by the American Association of State Highway and Transportation Officials
in Cooperation with the Federal Highway Administration

TRANSPORTATION RESEARCH BOARD

WASHINGTON, D.C.
2003
www.TRB.org

NATIONAL COOPERATIVE HIGHWAY RESEARCH PROGRAM

Systematic, well-designed research provides the most effective approach to the solution of many problems facing highway administrators and engineers. Often, highway problems are of local interest and can best be studied by highway departments individually or in cooperation with their state universities and others. However, the accelerating growth of highway transportation develops increasingly complex problems of wide interest to highway authorities. These problems are best studied through a coordinated program of cooperative research.

In recognition of these needs, the highway administrators of the American Association of State Highway and Transportation Officials initiated in 1962 an objective national highway research program employing modern scientific techniques. This program is supported on a continuing basis by funds from participating member states of the Association and it receives the full cooperation and support of the Federal Highway Administration, United States Department of Transportation.

The Transportation Research Board of the National Academies was requested by the Association to administer the research program because of the Board's recognized objectivity and understanding of modern research practices. The Board is uniquely suited for this purpose as it maintains an extensive committee structure from which authorities on any highway transportation subject may be drawn; it possesses avenues of communications and cooperation with federal, state and local governmental agencies, universities, and industry; its relationship to the National Research Council is an insurance of objectivity; it maintains a full-time research correlation staff of specialists in highway transportation matters to bring the findings of research directly to those who are in a position to use them.

The program is developed on the basis of research needs identified by chief administrators of the highway and transportation departments and by committees of AASHTO. Each year, specific areas of research needs to be included in the program are proposed to the National Research Council and the Board by the American Association of State Highway and Transportation Officials. Research projects to fulfill these needs are defined by the Board, and qualified research agencies are selected from those that have submitted proposals. Administration and surveillance of research contracts are the responsibilities of the National Research Council and the Transportation Research Board.

The needs for highway research are many, and the National Cooperative Highway Research Program can make significant contributions to the solution of highway transportation problems of mutual concern to many responsible groups. The program, however, is intended to complement rather than to substitute for or duplicate other highway research programs.

Project C12-48 FY'98

ISSN 0077-5614

ISBN 0-309-08750-3

Library of Congress Control Number 2003105418

© 2003 Transportation Research Board

Price $35.00

NOTICE

Published reports of the

NATIONAL COOPERATIVE HIGHWAY RESEARCH PROGRAM

are available from:

Transportation Research Board
Business Office
500 Fifth Street, NW
Washington, DC 20001

and can be ordered through the Internet at:

http://www.national-academies.org/trb/bookstore

Printed in the United States of America

THE NATIONAL ACADEMIES
Advisers to the Nation on Science, Engineering, and Medicine

AUTHOR ACKNOWLEDGMENTS

This report was written by Professor Michel Ghosn of the City College of the City University of New York/CUNY; Professor Fred Moses from the University of Pittsburgh; and Mr. Jian Wang, Research Assistant at the City College and the Graduate Center of CUNY. The authors would like to acknowledge the contributions of several participants and individuals who have contributed in various ways to the work presented in this report.

Dr. Roy Imbsen, Dr. David Liu, and Dr. Toorak Zokaie from Imbsen & Associates, Inc., helped with the earthquake models, provided bridge data, and assisted with project administration. Professor Peggy Johnson of Penn State University developed the scour reliability model. Professor George Mylonakis from CUNY contributed valuable input on foundation and S.S.I. modeling. Mr. Peter Buckland and Mr. Darrel Gagnon from Buckland & Taylor assisted in the Maysville Bridge analysis and provided foundation analysis models. Mr. Mark Hunter from Rowan, Williams, Davies, & Irwin,

Inc., performed the wind analysis of I-40 and Maysville bridges. Mr. Engin Aktas, Research Assistant at Pittsburgh, helped with the development of the live load model. Mr. Chuck Annis from Statistical Engineering, Inc., assisted in the statistical analysis of scour data.

Dr. Emil Simiu from NIST contributed input on wind load models. Dr. Arthur Frankel and E.V. Leyendecker from the U.S. Geological Survey provided the earthquake hazard data. Ms. Charlotte Cook and Ms. Debra Jackson from the U.S. Army Corps of Engineers provided the Mississippi vessel data. Mr. Henry Bollman, Florida DOT, contributed input on the vessel collision and scour models.

The authors are also especially grateful to the support and guidance provided by the project panel and the project director, Mr. David Beal. The project panel consisted of Mr. Thomas Post (Chair), Mr. Barry W. Bowers, Dr. Karen C. Chou, Mr. Christopher Dumas, Dr. Robert G. Easterling, Dr. Theodore V. Galambos, Dr. Jose Gomez, Dr. Vijayan Nair, Dr. Robert J. Perry, and Dr. Phil Yen.

FOREWORD

By David B. Beal
Staff Officer
Transportation Research
Board

NCHRP Report 489: Design of Highway Bridges for Extreme Events contains the findings of a study to develop a design procedure for application of extreme event loads and extreme event loading combinations to highway bridges. The report describes the research effort leading to the recommended procedure and discusses the application of reliability analysis to bridge design. The material in this report will be of immediate interest to bridge engineers and bridge-design specification writers.

The magnitude and consequences of extreme events such as vessel collisions, scour caused by flooding, winds, and earthquakes often govern the design of highway bridges. If these events are considered to occur simultaneously, the resulting loading condition may dominate the design. This superpositioning of extreme load values frequently increases construction costs unnecessarily because a simultaneous occurrence of two or more extreme events is unlikely. The reduced probability of simultaneous occurrence for each load combination may be determined using statistical procedures.

The *AASHTO LRFD Bridge Design Specifications* developed under NCHRP Project 12-33 cover the basic design combinations with dead load and live load. Extreme load combinations were not considered in the load resistance factor design (LRFD) calibration because of the lack of readily available data concerning the correlation of extreme events. Nevertheless, a probability-based approach to bridge design for extreme events can be accomplished through incorporation of state-of-the-art reliability methodologies.

The objective of NCHRP Project 12-48 was to develop a design procedure for the application of extreme event loads and extreme event loading combinations to highway bridges. This objective has been achieved with a recommended design procedure consistent with the uniform reliability methodologies and philosophy included in the *AASHTO LRFD Bridge Design Specifications*. Four new extreme event load combinations are included to maintain a consistent level of safety against failure caused by scour combined with live load, wind load, vessel collision, and earthquake, respectively.

This research was performed at the City College of the City University of New York, with the assistance of Dr. Fred Moses. The report fully documents the methodology used to develop the extreme load combinations and the associated load factors. Recommended specification language is included in a published appendix. All appendixes to the report are included on *CRP-CD-30*.

CONTENTS

1   **SUMMARY**

7   **CHAPTER 1   Introduction**
        1.1   Load Combinations in Current Codes, 7
        1.2   Combination of Extreme Events for Highway Bridges, 9
        1.3   Reliability Methods for Combination of Extreme Load Effects, 10
        1.4   Reliability-Based Calibration of Load Factors, 13
        1.5   Research Approach, 15
        1.6   Report Outline, 16

17   **CHAPTER 2   Reliability Models for Combinations of Extreme Events**
        2.1   Loads and Return Periods in AASHTO LRFD, 17
        2.2   Basic Concepts of Structural Reliability, 20
        2.3   Resistance Models, 23
        2.4   Load Models, 27
        2.5   Risk Assessment Models for Load Combinations, 45
        2.6   Chapter Conclusions, 49

50   **CHAPTER 3   Calibration of Load Factors for Combinations of Extreme Events**
        3.1   Description of Basic Bridge Configurations and Structural Properties, 50
        3.2   Reliability Analysis for Extreme Events, 60
        3.3   Reliability Analysis for Combinations of Extreme Events and
               Calibration of Load Factors, 77
        3.4   Summary and Recommendations, 96

100   **CHAPTER 4   Conclusions and Future Research**
        4.1   Conclusions, 100
        4.2   Future Research, 102

106   **REFERENCES**

108   **GLOSSARY OF NOTATIONS**

110   **APPENDIXES**
        Introduction, 110
        Appendix A: Recommended Modifications to AASHTO
          LRFD Bridge Design Specifications, 111
        Appendix B: Reliability Model for Scour Analysis, 125
        Appendix C: Reliability Analysis of Three-Span Bridge Model, 128
        Appendix H: Seismic Risk Analysis of a Multispan Bridge, 152
        Appendix I: Analysis of Scour Data and Modified Reliability
          Model for Scour, 162

DESIGN OF HIGHWAY BRIDGES
FOR EXTREME EVENTS

SUMMARY   The current AASHTO load resistance factor design (LRFD) specifications were developed using a reliability-based calibration that covered gravity loads consisting of the basic combination of permanent (or dead) load plus live load. The other load combinations were obtained from previous generations of specifications and from the experience of bridge engineers and, thus, may not be consistent with the reliability methodology of the LRFD specifications. The objective of this study is to develop a design procedure for the consideration of extreme events and the combination of their load effects in the *AASHTO LRFD Bridge Design Specifications*. Extreme events are defined as man-made or environmental hazards having a high potential for producing structural damage but are associated with a relatively low rate of occurrence. The extreme events considered in this study include live loads, earthquakes, wind loads, ship collision forces, and scour.

According to the AASHTO LRFD, bridges should be designed for a 75-year return period. The probability that a bridge will be subjected in its 75-year design life to an extreme event of a certain magnitude depends on the rate of occurrence of the event and the probability distribution of the event's intensity. Generally speaking, there is low probability that several extreme events will occur simultaneously at any point in time within a bridge's design life. Even when simultaneous occurrences do occur, the chances that all the events are at their highest intensities are very small. To account for these low probabilities, engineers have historically used the one-third stress-reduction rule when combining extreme environmental events, such as wind or earthquake loads, with gravity loads. This rule, which dates back to the early years of the 20th century, has been discredited; it is generally accepted that a more appropriate procedure should use load combination factors derived from the theory of structural reliability.

The aim of structural reliability theory is to account for the uncertainties encountered during the safety evaluation of structural systems or during the calibration of load and resistance factors for structural design codes. The uncertainties considered include those associated with predicting the load-carrying capacity of a structure, the intensities of the extreme events expected to be applied during the structure's design life, the frequency of these loading events, and the prediction of the effects of these events on the structure.

To ensure the safety of highway bridges under the combined effects of extreme events, this study develops load factors appropriate for inclusion in the AASHTO LRFD design-check equations. The reliability analysis of the effects of each threat taken individually is performed using methods developed in previous bridge code calibration efforts (for the live loads and ship collisions) and during the development of other structural codes (for wind loads and earthquake loads). Because the current AASHTO specifications for scour are not based on reliability methods, a scour reliability model is developed for the purposes of this study. Results of reliability analysis of typical bridge configurations under the effect of individual threats are used to define target reliability levels for the development of load factors applicable for designing bridges that may be susceptible to combinations of threats.

To achieve the objectives of the study, this project first reviews the basic reliability methodology used during previous code calibration efforts. Basic bridge configurations designed to satisfy the current AASHTO specifications are analyzed to find the implicit reliability index values for different limit states when the bridges are subjected to live loads, wind loads, earthquakes, vessel collisions, or scour. The limit states considered include column bending, shearing failure, axial failure of bridge columns, bearing failure of column foundations, and overtipping of single-column bents. The reliability analysis uses appropriate statistical data on load occurrences and load intensities for the pertinent extreme events that are assembled from the reliability literature and United States Geological Survey (USGS) websites. Statistical data on member and foundation capacities and load analysis models commonly used in reliability-based code calibration efforts are also used. Reliability indexes are calculated for the same bridges when subjected to combinations of extreme events using the Ferry-Borges model. The results are subsequently used to calibrate load combination factors appropriate for implementation in the LRFD equations.

The Ferry-Borges model assumes that each extreme event type produces a sequence of independent load effects, each lasting for an equal duration of time. The service life of the structure is then divided into equal intervals of time. The probability that a load occurs in an arbitrary time interval can be calculated from the event's occurrence rate. Simultaneously, the probability distribution of the intensity of the load given that the event has occurred can be calculated from statistical information on load intensities. The probability that a second event would occur in the same time interval when the first load event is on can also be calculated from the rate of occurrence of the second load and the time durations of each load. After calculating the probability density for the second load given that it has occurred, the probability of the intensity of the combined load effects can be calculated using a convolution integral. The load combination problem consists of predicting the maximum value of the combined load effects that is likely to occur in the lifetime of the bridge. Although the Ferry-Borges model gives a simplified representation of the actual loading phenomenon, this model is more accurate than other load combination rules such as Turkstra's rule because it takes into consideration the rate of occurrence of the loads and their time durations. The probability distribution of the maximum value of the combined load effect is used along with statistical data on bridge member and system resistances to find the probability of failure and the reliability index, β.

The load factors proposed in this study are calibrated such that bridges subjected to a combination of events provide reliability levels similar to those of bridges with the same configurations but situated in sites where one threat is dominant. Thus, the proposed load factors are based on previous experiences with "safe bridge structures" and provide balanced levels of safety for each load combination. The results of this study indicate that different threats produce different reliability levels; therefore, the target reliability indexes for the combination of events are selected in most cases to provide the same reliability level associated with the occurrence of the individual threat with the highest reliability index. Thus, when dealing with the combination of live load plus wind load or live load plus scour, the reliability index associated with live loads is used as target. When studying the reliability of bridges subjected to the combination of wind loads and scour, the reliability index associated with wind loads alone is chosen for target. Similarly, when studying the reliability of vessel collision with scour or vessel collision with wind load, the reliability index associated with vessel collisions is used for target. For combinations involving earthquake loads, it is the reliability index associated with earthquakes alone that is used for target even if the reliability for earthquakes alone produces a lower reliability index. Combinations involving earthquakes are treated differently than other combinations because of the large additional capacity and resulting construction costs that would be required to increase the reliability levels of bridges subjected to earthquake risks.

The analysis considers structural safety as well as foundation safety. For multicolumn bents, system safety is compared with member safety. The results show that the system produces a reliability index about 0.25 higher than the reliability index of the individual members for two-column bents formed by unconfined concrete columns. Hence, the system factors calibrated under NCHRP Project 12-47 are applicable for the cases in which linear elastic analysis is performed to check bridge member safety (see *NCHRP Report 458: Redundancy in Highway Bridge Substructures* [Liu et al., 2001]). NCHRP Project 12-47 calibrated system factors for application on the left-hand side of the design equation to complement the member resistance factor. The cases for which the application of system factors is possible include the analysis of bridges subjected to combinations exclusively involving live loads, wind loads, and ship collision forces. The analysis for combinations involving earthquakes is based on the plastic behavior of bridge bents; thus, system safety is directly considered and no system factors need to be applied. Scour causes the complete loss of the load-carrying capacity of a column, and bridge bents subjected to scour depths exceeding the foundation depth will have little redundancy. Thus, such failures should be associated with system factors on the order of 0.80 as recommended by NCHRP Project 12-47.

Results of the reliability analyses indicate that there are large discrepancies among the reliability levels implied in current design practices for the different extreme events under consideration. Specifically, the following observations are made:

- The AASHTO LRFD was calibrated to satisfy a target member reliability index equal to 3.5 for gravity loads. The calculations performed in this study confirm that bridge column bents provide reliability index values close to the target 3.5 for the different limit states considered. These limit states include column bending and axial failure for one-column and multicolumn bents, as well as overtipping of one-column bents. Bearing failure of the soil may produce lower reliability levels depending on the foundation analysis model used.
- The system reliability index for bridge bents subjected to earthquakes is found to be on the order of 2.9 for moment capacity or 2.4 for overtipping of single-column bents founded on pile extensions (drilled shafts) that can be inspected. Lower reliability index values are observed for other subsystems depending on the response modification factors used during the design of their components. Unlike the analysis for other hazards, the earthquake analysis procedure accounts for system capacity rather than for member capacity because the earthquake analysis process accounts for plastic redistribution of loads and failure is defined as a function of the ductility capacity of the members. Although this is relatively low compared with the

4

member reliability index for gravity loads, the engineering community is generally satisfied with the safety levels associated with current earthquake design procedures, and increases in the currently observed safety levels would entail high economic costs. For this reason, the target reliability index for load combination cases involving earthquakes is chosen to be the same reliability index calculated for designs satisfying the current design criteria when earthquakes alone are applied. On the other hand, a future review of the response modification factors used in earthquake design is recommended in order to produce more uniform reliability levels for all system types.

- The reliability index for designing bridge piers for scour in small rivers varies from about 0.45 to 1.8, depending on the size of the river and the depth and speed of the discharge flow. These values are much lower than the 3.5 target for gravity loads and are also lower than the index values observed for earthquakes. In addition, failures caused by scour may often lead to total collapse as compared with failures of members under gravity loads. Therefore, it is recommended to increase the reliability index for scour by applying a scour safety factor equal to 2.0. The application of the recommended 2.0 safety factor means that if current HEC-18 scour design procedures are followed, the final depth of the foundation should be 2.0 times the value calculated using the HEC-18 equation. Such a safety factor will increase the reliability index for scour from an average of about 1.0 for small rivers to a value slightly higher than 3.0, which will make the scour design safety levels compatible with the safety levels for other threats. However, a review of the HEC-18 equations is recommended in order to provide more uniform safety levels for all river categories.
- While bridge design methods for wind loads provide an average member reliability index close to 3.0, there are large differences among the reliability indexes obtained for different U.S. sites. For this reason, it is recommended that future research in wind engineering develop new wind design maps that would provide more uniform safety levels for different regions of the United States.
- The AASHTO vessel collision model produces a reliability index of about 3.15 for shearing failures and on the order of 2.80 for bending failures. The presence of system redundancy caused by the additional bending moment resistance by the bents, abutments, or both that are not impacted would increase the reliability index for bending failures to more than 3.00, making the safety levels more in line with those for shearing failures.

The recommended load combination factors are summarized in Appendix A in a format that is implementable in the AASHTO LRFD specifications. The results illustrate the following points:

- The current load factors for the combination of wind plus live loads lead to lower reliability indexes than do those of either load taken separately. Hence, this study has recommended increasing the load factors for wind on structures and wind on live loads from the current 0.40 to 1.20 in combination with a live load factor of 1.0 (instead of the current live load factor of 1.35).
- The commonly used live load factor equal to 0.50 in combination with earthquake effects would lead to conservative results. This report has shown that a load factor of 0.25 on live load effects when they are combined with earthquake effects would still provide adequate safety levels for typical bridge configurations subjected to earthquake intensities similar to those observed on either the west or east coasts. These calculations are based on conservative assumptions on the recurrence of live loads when earthquakes are actively vibrating the bridge system.

- For the combination of vessel collision forces and wind loads, a wind load factor equal to 0.30 is recommended in combination with a vessel collision factor of 1.0. The low wind load factor associated with vessel collisions compared with that recommended for the combination of wind loads plus live loads partially reflects the lower rate of collisions in the 75-year design life of bridges as compared with the number of live load events.
- A scour factor equal to 1.80 is recommended for use in combination with a live load factor equal to 1.75. The lower scour load factor for the combination of scour and live loads reflects the lower probability of having the maximum possible 75-year live load occur when the scour erosion is also at its maximum 75-year depth.
- A scour factor equal to 0.70 is recommended in combination with a wind load factor equal to 1.40. The lower scour factor observed in combination with wind loads as compared with the combination with live loads reflects the lower number of wind storms expected in the 75-year design life of the structure.
- A scour factor equal to 0.60 is recommended in combination with vessel collision forces. The lower scour factor observed in combination with collision forces reflects the lower number of collisions expected in the 75-year bridge design life.
- A scour factor equal to 0.25 is recommended in combination with earthquakes. The lower scour factor with earthquakes reflects the fact that, as long as a total washout of the foundation does not occur, bridge columns subjected to scour exhibit lower flexibilities that will help reduce the inertial forces caused by earthquakes. This reduction in inertial forces partially offsets the scour-induced reduction in soil depth and the resulting soil-resisting capacity.

With regard to the extreme loads of interest to this study, the recommended revisions to the *AASHTO LRFD Bridge Design Specifications* (1998) would address the loads by ensuring that the factored member resistances are greater than the maximum load effects obtained from the following combinations (see following paragraphs for variable definition):

- Strength I Limit State: $1.25\,DC + 1.75\,LL$
- Strength III Limit State: $1.25\,DC + 1.40\,WS$
- Strength V Limit State: $1.25\,DC + 1.00\,LL + 1.20\,WS + 1.20\,WL$
- Extreme Event I: $1.25\,DC + 0.25\,LL + 1.00\,EQ$
- Extreme Event II: $1.25\,DC + 0.25\,LL + 1.00\,CV$, or
  $1.25\,DC + 0.30\,WS + 1.00\,CV$
- Extreme Event III: $1.25\,DC;\ 2.00\,SC$, or
  $1.25\,DC + 1.75\,LL;\ 1.80\,SC$
- Extreme Event IV: $1.25\,DC + 1.40\,WS;\ 0.70\,SC$
- Extreme Event V: $1.25\,DC + 1.00\,CV;\ 0.60\,SC$
- Extreme Event VI: $1.25\,DC + 1.00\,EQ;\ 0.25\,SC$

The presence of scour is represented by the variable *SC*. The semicolon indicates that the analysis for load effects should assume that a maximum scour depth equal to $\gamma_{sc}\,SC$ exists when the other load events are applied where *SC* is the scour depth calculated from the HEC-18 equations. When scour is possible, the bridge foundation should always be checked to ensure that the foundation depth exceeds 2.00 *SC*. For the cases involving a dynamic analysis such as the analysis for earthquakes, it is critical that the case of zero scour depth be checked because in many cases, the presence of scour may reduce the applied inertial forces. The resistance factors depend on the limit states being considered. When a linear elastic analysis of single and multicolumn

6

bents is used, the system factors developed under NCHRP Project 12-47 should also be applied.

In the equations given above, *DC* represents the dead load effect, *LL* is the live load effect, *WS* is the wind load effect on the structure, *WL* is the wind load acting on the live load, *EQ* is the earthquake forces, *CV* is the vessel collision load, and *SC* represents the design scour depth. The dead load factor of 1.25 would be changed to 0.9 if the dead load counteracts the effects of the other loads.

The recommended changes in the AASHTO LRFD consist of adding Extreme Event Cases III through VI, which consider scour. In addition, Extreme Event II is modified to include a check of either live loads or wind loads with vessel collision forces. A higher wind load factor than live load factor is used to reflect the fact that the rate of vessel collisions increases during the occurrence of windstorms.

CHAPTER 1

INTRODUCTION

This study is concerned with the safety of bridges subjected to the combination of four types of extreme load events: (1) earthquakes, (2) winds, (3) scour, and (4) ship collisions. In addition, these loads will combine with the effects of truck live loads and the effects of dead loads. To ensure the safety of highway bridges under the combined effects of extreme events, this study develops load factors appropriate for inclusion in the AASHTO LRFD design-check equations. Structural reliability methods are used during the load factor calibration process in order to be consistent with the LRFD philosophy and to account for the large uncertainties associated with the occurrence of such extreme events, estimating their intensities, and analyzing their effects on bridge structures.

Analytical models to study the probability of single events and multiple load occurrences are available in the reliability literature and have been used to calibrate a variety of structural design codes ranging from buildings, to offshore platforms, to nuclear power plants, to transmission towers, to ships. This chapter describes the research approach followed during the course of this study, provides an overview of how current codes consider load combinations of extreme events, and describes how reliability analysis methods can be used to calibrate load factors.

1.1 LOAD COMBINATIONS IN CURRENT CODES

Historically, engineers used the one-third stress-reduction rule when combining extreme environmental events, such as wind or earthquake loads, with gravity loads. This rule, which dates to the early 20th century, has been discredited and replaced by load combination factors derived from reliability analyses. The reliability-based effort on load combinations has eventually lead to the development of the American National Standards Institute (ANSI) A58 Standard (ANSI A58 by Ellingwood et. al., 1980). The ANSI document set the stage for the adoption of similar load combination factors in current generation of structural design codes such as the *Manual of Steel Construction* (American Institute of Steel Construction [AISC], 1994), also called the *AISC Manual*; ACI 318-95 (American Concrete Institute [ACI], 1995); *AASHTO LRFD Bridge Design Specifications* (AASHTO, 1998); and many other codes.

The total number of load combinations covered in *AASHTO LRFD Bridge Design Specifications* includes five combinations to study the safety of bridge members for strength limit states, two load combinations for extreme load events, three load combinations for service load conditions, and finally the fatigue loading. The loads in these combinations include the effects of the permanent loads (which cover the weights of the structural components, attachments, and wearing surface) and earth pressure. The transient loads include those caused by the motion of vehicles and pedestrians; environmental effects such as temperature, shrinkage, and creep; water pressure; the effect of settlements and foundations; ice; collision; wind; and earthquakes.

With regard to the extreme loads of interest to this study, the current version of the AASHTO LRFD specifications (AASHTO, 1998) addresses them through the following combinations:

- Strength I Limit State: $1.25\,DC + 1.75\,LL$
- Strength III Limit State: $1.25\,DC + 1.40\,WS$
- Strength V Limit State: $1.25\,DC + 1.35\,LL + 0.4\,WS + 0.4\,WL$ (1.1)
- Extreme Event I: $1.25\,DC + 1.00\,EQ$
- Extreme Event II: $1.25\,DC + 0.50\,LL + 1.00\,CV$

DC is the dead load, LL is the live load, WS is the wind load on structure, WL is the wind load acting on the live load, EQ is the earthquake load, and CV is the vessel collision load. The dead load factor of 1.25 would be changed to 0.9 if the dead load counteracts the effects of the other loads.

The safety check involving EQ is defined in the AASHTO specifications as "Extreme Event I, Limit State." It covers earthquakes in combination with the dead load and a fraction of live load to "be determined on a project-specific basis," although the commentary recommends a load factor equal to 0.5. "Extreme Event II" considers 50% of the design live load in combination with either ice load, vessel collision load, or truck collision load. Wind loads are considered in "Strength Limit III," in which they do not combine with the live load, and in "Strength Limit V," in which they do combine with the live load. It should be noted that the design (nominal) live load for the strength limit states was based on the maximum 75-year truck weight combination and as such

8

should also be considered an extreme event situation. In most strength limit states, it is the live load factor that changes from 1.75 down to possibly 0.5 when the other extreme loads are associated with a load factor of 1.0. The one exception is the wind load: when used alone with the dead load, it is associated with a 1.4 load factor and a 0.4 factor when combined with the live load. The AASHTO LRFD specifications do not consider the possibility of combining wind, earthquake, and vessel collision, nor do they account for the effect of scour-weakened foundations when studying the safety under any of the transient loads. Also, as mentioned above, the LRFD commentary suggests that 0.50 *LL* may be added to the case when the earthquake load is acting, although the specifications state that "it should be determined on a project-specific basis." The nominal values for the loads correspond to different return periods depending on the extreme event. The return periods vary widely from 50 years for wind, to 75 years for live load, to 2,500 years for earthquakes, while the ship collision design is decided based upon an annual failure rate. Although not specifically addressed in the AASHTO LRFD specifications, the design for scour as currently applied based on FHWA HEC-18 models uses a 100-year return period as the basis for safety evaluation. It is noted that for most load events, the design nominal values and return periods are associated with implicit biases that would in effect produce return periods different than those "specified." For example, the HL-93 design live load is found to be smaller than the 75-year maximum live load calculated from the data collected by Nowak (1999) (see *NCHRP Report 368: Calibration of LRFD Bridge Design Code*) by about a factor of about 1.20. Similarly, the wind maps provided by ASCE 7-95 and adopted by the AASHTO LRFD specifications give a biased "safe" envelope to the projected 50-year wind speeds. These issues are further discussed in Chapter 2.

The load factors and load combinations specified in the current version of the AASHTO LRFD specifications follow the same trends set by other structural codes that were calibrated using reliability-based methods. For example, AISC's *Manual of Steel Construction* bases its load combinations on the provisions of ANSI A58. These provisions account for the dead load, live load (roof live load is treated separately), wind load, snow load, earthquake load, ice load, and flooding. The primary load combinations are 1.2 dead load plus 1.6 live load and 0.5 of snow or ice or rain load, another combination considers 1.2 dead load plus 1.6 snow or ice plus 0.5 live or 0.8 wind load. The 1.2 dead load is also added to 1.5 times the nominal earthquake load plus 0.5 live load or 0.2 times the snow load.

ANSI A58 has eventually been replaced by the ASCE 7-95 document (American Society of Civil Engineers [ASCE], 1995). The latter gives a long list of load combinations, some of which have been modified from earlier versions. For example, the primary combination for seismic load includes 1.2 times the dead load, 1.0 times the earthquake load, and 0.5 times the live load plus 0.2 times the snow load. This com-

bination uses a seismic load factor of 1.0 based on the National Earthquake Hazards Reduction Program recommendations (NEHRP, 1997) rather than on the 1.5 initially proposed in the ANSI code. The magnitudes of the earthquake forces are based on ground accelerations with 10% probability of being exceeded in 50 years. This is equivalent to a 475-year return period. Recent work by NEHRP has supported increasing the earthquake return period up to 2,500 years, although the latest recommendation allows for a two-thirds reduction in the earthquake magnitudes for certain cases. The return period for the wind loads is set at 50 years.

The ACI Building Code's primary load combination is 1.4 dead load plus 1.7 live load. When wind is considered, the code includes a 1.7 wind load factor but then reduces the total load by 25%. Earth pressure is treated in the same manner as the live load. When fluid pressure is included, it is associated with a load factor of 1.4 and then added to 1.4 dead load and 1.7 live load. Loads due to settlement, creep, temperature, and shrinkage are associated with a load factor of 1.4 and when added to dead and live load, a 25% reduction in the total load is stipulated.

Other agencies that have implemented specifications for load combinations include the U.S. Nuclear Regulatory Commission (US-NRC) and the American Petroleum Institute (API). The US-NRC *Standard Review Plan* (US-NRC, 1989) for nuclear power plants uses load combinations similar to those provided in ASCE 7-95. Extreme event loads are each treated separately with a load factor of 1.0. The only difference between the US-NRC and the ASCE 7-95 provisions are the higher return periods imposed on the extreme event loads, particularly the seismic accelerations.

1.1.1 Summary

All of the codes listed above were developed using reliability-based calibrations of the load factors; yet, it is observed that the types of loads that are considered simultaneously and the corresponding load factors differ considerably from code to code. Some of the differences in the load factors may be justified because of variations in the relative magnitudes of the dead loads and the transient loads for the different types of structures considered (e.g., dead load to live load ratio). Also, in addition to assigning the load factors, an important component of the specifications is the stipulation of the magnitude of the design loads or the return period for each load. For example, if the design wind load corresponds to the 50-year storm with a load factor of 1.4, it may produce a similar safety level as the 75-year storm with a load factor of 1.0. One should also account for the hidden biases and conservativeness built into the wind maps and other load data. Finally, one should note that the safety level is related to the ratio of the load factor to the resistance factor. For example, if the load factor were set at 1.40 with a resistance factor of 1.0, it would produce a similar safety level as a load factor of 1.25 when associated with a resistance factor of 0.90.

Despite the justifications for the differences mentioned above, variations in the load factors may still lead to differences in the respective safety levels implicit in each code. These differences are mainly due to the nature of the currently used reliability-based calibration process, which uses a notional measure of risk rather than an actuarial value. In most instances, the code writers propose (1) load return periods and (2) resistance and load factors based on a "calibration" with past experience to ensure that new designs provide the *same* "safety levels" as existing structures judged to be "acceptably safe." Because the historical evolution of the AASHTO, ACI, AISC, and the other codes may have followed different paths, it is not surprising to see that the calibration process produces different load combinations and different load factors. In addition, most codes use the reliability of individual members as the basis for the calibration process. However, different types of structures have different levels of reserve strength such that for highly redundant structures, the failure of one member will not necessarily lead to the collapse of the whole structure. Therefore, in many cases, the actual reliability of the system is significantly higher than the implied target reliability used during the calibration process. Even within one system, the reliability levels of subsystems may differ—for example, for bridges, the superstructure normally formed by multiple girders in parallel would have higher system reserve than would the substructure, particularly when the substructure is formed by single-column bents.

In particular, the AASHTO LRFD specifications were developed using a reliability-based calibration that covered only dead and live loads (Strength I Limit State) using a target reliability index equal to 3.5 against first member failure. The other load combinations were specified based on AASHTO's *Standard Specifications for Highway Bridges* (AASHTO, 1996); on common practice in bridge engineering; or on the results obtained from other codes. Consequently, the current provisions in the AASHTO LRFD for load combinations may not be consistent with the provisions of the Strength I Limit State and may not produce consistent safety levels, as was the original intent of the specification writers.

In theory, when looking at the possibility of load combinations, an infinite number of combinations are possible. For example, the maximum combined live load and earthquake load effect might occur with the largest earthquake, or the second largest earthquake, and so forth, depending on the contribution of the live load in each case to the total effect. The purpose of the calibration process is to provide a set of design loads (or return periods) associated with appropriate load factors to provide an "acceptably safe" envelope to all these possible combinations. The term "acceptably safe" is used because absolute safety is impossible to achieve. Also, there is a trade-off between safety and cost. The safer the structure is designed to be, the more expensive it will be to build. Hence, code writers must determine how much implicit cost they are willing to invest to build structures with extremely high levels of safety.

The next section will review available methods to study the reliability of structures subjected to the combination of load events. This information will be essential to determine how and when extreme load events will combine and what load factors will provide a safe envelope to the risk of bridge failure due to individual load events and the combination of events.

1.2 COMBINATION OF EXTREME EVENTS FOR HIGHWAY BRIDGES

The extreme events of concern to this project are transient loads with relatively low rates of occurrences and uncertain intensity levels. Once an extreme event occurs, its time duration is also a random variable with varying length, depending on the nature of the event. For example, truck loading events are normally of very short duration (on the order of a fraction of a second to 2 to 3 seconds) depending on the length of the bridge, the speed of traffic, and the number of trucks crossing the bridge simultaneously (platoons). Windstorms have varying ranges of time duration and may last for a few hours. Most earthquakes last for 10 to 15 sec while ship collisions are instantaneous events. On the other hand, the effects of scour may last for a few months for live bed scour and for the remainder of the life of a bridge pier for clear water conditions. The transient nature of these loads, their low rate of occurrence, and their varying duration times imply that the probability of the simultaneous occurrence of two events is generally small. The exceptions are when one of the loads occurs frequently (e.g., truckloads); when the two loads are correlated (ship collision and windstorm); or when one of the loads lasts for long time periods (scour or, to a lesser extent, wind).

Even when two load types occur simultaneously, there is little chance that the intensities of both events will be close to their maximum lifetime values. For example, the chances are very low that the trucks crossing a bridge are very heavily loaded at the time of the occurrence of a high-velocity windstorm. On the other hand, because ship collisions are more likely to occur during a windstorm, the effect of high wind velocities may well combine with high-impact loads from ship collisions. Also, once a bridge's pier foundations have been weakened because of the occurrence of scour, the bridge would be exposed to high risks of failure because of the occurrence of any other extreme event. Of the extreme events of interest to this study, only ship collision and wind speeds are correlated events. Although scour occurs because of floods that may follow heavy windstorms, the time lag between the occurrence of a flood after the storm would justify assuming independence between wind and scour events.

For the purposes of this study and following current practice, it will be conservatively assumed that the intensity of any extreme event will remain constant at its peak value for the time duration of the event. The time duration of each event will be assumed to be a pre-set deterministic constant value. The occurrence of extreme load events may be represented as depicted in Figure 1.1, which shows how the intensities may be modeled as constant in time once the event occurs although the actual intensities generally vary with time.

Methods to study the combinations of the effects of extreme events on structural systems have been developed based on the theory of structural reliability. Specifically, three analytical models for studying the reliability of structures under the effect of combined loads have been used in practical applications. These are (1) Turkstra's rule; (2) the Ferry-Borges (or Ferry Borges–Castanheta) model; and (3) Wen's load coincidence method. In addition, simulation techniques such as Monte Carlo simulations are applicable for any risk analysis study. These methods are intended to calculate the probability of failure of a structure subjected to several transient loads and have been used to calibrate a variety of structural codes ranging from bridges, to buildings, to offshore platforms, to nuclear power plants, to transmission towers, to ships. The next section describes the background and the applicability of these methods.

1.3 RELIABILITY METHODS FOR COMBINATION OF EXTREME LOAD EFFECTS

Early structural design specifications represented the load combination problem in a blanket manner by simply decreasing the combined load effect of extreme events by 25% (e.g., ACI) or by increasing the allowable stress by 33% (e.g., AISC–allowable stress design [ASD]). These approaches do not account for the different levels of uncertainties associated with each of the loads considered, nor do they consider the respective rates of load occurrence and duration. For example, these methods decrease the dead load effect by the same percentage as the transient load effects although the dead

load is normally better known (has a low level of uncertainty), is always present, and remains constant with time. The use of different load factors depending on the probability of simultaneous occurrences of the loads is generally accepted as the most appropriate approach that must be adopted by codes in dealing with the combination of loads.

An accurate calibration of the load factors for the combination of extreme loads requires a thorough analysis of the fluctuation of loads and load effects during the service life of the structure. The fluctuations of the load effects in time can be modeled as random processes (as illustrated in Figure 1.1) and the probability of failure of the structure can be analyzed by studying the probability that the process exceeds a threshold value corresponding to the limit state under consideration. Each loading event can be represented by its rate of occurrence in time, its time duration, and the intensity of the load. In addition, for loads that produce dynamic responses, the effects of the dynamic oscillations are needed. Several methods of various degrees of accuracy and simplicity are available to solve this problem. Three particular methods have been used in the past by different code-writing groups. These are (1) Turkstra's rule, (2) the Ferry Borges–Castanheta model, and (3) Wen's load coincidence method (for examples, see Thoft-Christensen and Baker, 1982; Turkstra and Madsen, 1980; or Wen, 1977 and 1981). These load combination models can be included in traditional first order reliability method (FORM) programs. In addition, Monte Carlo simulations can be used either to verify the validity of the models used or to directly perform the reliability analysis. Results of the reliability analysis will be used to (1) select the target reliability levels and (2) verify that the selected load factors would produce designs that uniformly satisfy the target reliability levels. Below is a brief description of the three analytical methods. Chapter 2 will provide a more detailed

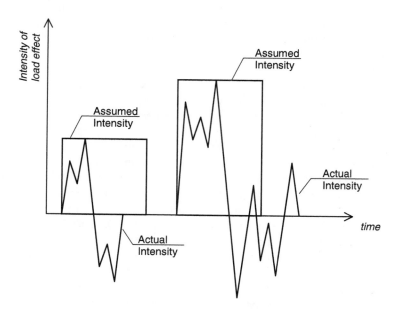

Figure 1.1. Modeling the effect of transient loads.

description of the method used in this study. Chapter 3 and the appendixes provide illustrations on the application of the selected method of analysis and the results obtained.

1.3.1 Turkstra's Rule

Turkstra's rule (Turkstra and Madsen, 1980) is a deterministic (non-random) procedure to formulate a load combination format for the design of structures subjected to the combined effects of several possible loading events. The rule is an over-simplification derived from the more advanced Ferry Borges–Castanheta model. Assuming two load types only (e.g., live load plus wind load), the intensity of the effect of Load 1 is labeled as x_1 and for Load 2, the intensity is defined as x_2. Both x_1 and x_2 are random variables that vary with time. Turkstra's rule may be summarized as follows:

- Design for the largest lifetime maximum value of Load 1 plus the value of Load 2 that will occur when the maximum value of Load 1 is on.
- Also design for the lifetime maximum of Load 2 plus the value of Load 1 that will occur when Load 2 is on.
- Select the larger of these two designs.

In practical situations, the value of the load that is not at its maximum is taken at its mean (or expected) value. Turkstra's rule can thus be expressed as follows:

$$X_{\text{max},T} = \max \left\{ \begin{matrix} [\max(x_1) + \bar{x}_2] \\ [\bar{x}_1 + \max(x_2)] \end{matrix} \right\} \tag{1.2}$$

where

$X_{\text{max},T}$ = the maximum value for the combined load effects in a period of time T,
$\max(x_1)$ = the maximum of all possible x_1 values,
$\max(x_2)$ = the maximum of all possible x_2 values,
$\bar{x}_1$ = the mean value of x_1, and
$\bar{x}_2$ = the mean value of x_2.

The rule can be extended for more than two loads following the same logic. Although simple to use, Turkstra's rule is generally found to provide inconsistent results and is often unconservative (Melchers, 1999).

1.3.2 Ferry Borges–Castanheta Model for Load Combination

The Ferry Borges–Castanheta model is herein described for two load processes and illustrated in Figure 1.2 (Turkstra and Madsen, 1980; Thoft-Christensen and Baker, 1982). The model assumes that each load effect is formed by a sequence of independent load events, each with an equal duration. The service life of the structure is then divided into equal inter-

vals of time, each interval being equal to the time duration of Load 1, t_1. The probability of Load 1 occurring in an arbitrary time interval can be calculated from the occurrence rate of the load. Simultaneously, the probability distribution of the intensity of Load 1 given that the load has occurred can be calculated from statistical information on load intensities. The probability of Load 2 occurring in the same time interval as Load 1 is calculated from the rate of occurrence of Load 2 and the time duration of Loads 1 and 2. After calculating the probability density for Load 2 given that it has occurred, the probability of the intensity of the combined loads can be easily calculated.

The load combination problem consists of predicting the maximum value of the combined load effect X, namely $X_{\text{max},T}$, that is likely to occur in the lifetime of the bridge, T. In the lifetime of the bridge there will be n_1 independent occurrences of the combined load, X. The maximum value of the n_1 possible outcomes is represent by

$$X_{\text{max},T} = \max_{n_1}[X] \tag{1.3}$$

The maximum value of x_2 that is likely to occur within a time period t_1 (i.e., when Load 1 is on) is defined as $x_{2\,\text{max},t1}$. Since Load 2 occurs a total of n_2 times within the time period t_1, $x_{2\,\text{max},t1}$ is represented by

$$x_{2\,\text{max},t1} = \max_{n_2}[x_2] \tag{1.4}$$

$X_{\text{max},T}$ can then be expressed as

$$X_{\text{max},T} = \max_{n_1}[x_1 + x_{2\,\text{max},t1}] \tag{1.5}$$

or

$$X_{\text{max},T} = \max_{n_1}\left[x_1 + \max_{n_2}(x_2)\right]. \tag{1.6}$$

The problem reduces then to finding the maximum of n_2 occurrences of Load 2, adding the effect of this maximum to the effect of Load 1, then taking the maximum of n_1 occurrences of the combined effect of x_1 and the n_2 maximum of Load 2. This approach assumes that x_1 and x_2 have constant intensities during the duration of one of their occurrences. Notice that x_1 or x_2 could possibly have magnitudes equal to zero. If the intensities of x_1 and x_2 are random variables with known probability distribution functions, then the probability distribution functions of the maximum of several events can be calculated using Equation 1.7.

The cumulative distribution of a single load event, Y, can be represented as $F_Y(Y^*)$. $F_Y(Y^*)$ gives the probability that the variable Y takes a value less than or equal to Y^*. Most load combination studies assume that the load intensities are independent from one occurrence to the other. In this case, the cumulative distribution of the maximum of m events that

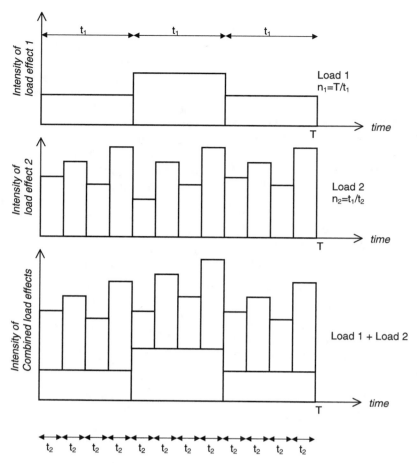

Figure 1.2.   Illustration of combination of two load effects according to Ferry Borges–Castanheta model.

occur in a time period T can be calculated from the probability distribution of one event by

$$F_{Y_{\max},\,m}(Y^*) = F_Y(Y^*)^m \qquad (1.7)$$

where m is the number of times the load Y occurs in the time period T.

Equation 1.7 is obtained by realizing that the probability that the maximum value of m occurrences of load Y is less than or equal to Y^* if the first occurrence is less than or equal to Y^*, *and* the second occurrence is less than or equal to Y^*, *and* the third occurrence is less than or equal to Y^*, and so forth. This is repeated m times, which leads to the exponent, m, in the right-hand-side term of Equation 1.7.

This approach, which assumes independence between the different load occurrences, has been widely used in many previous efforts of calibration of load factors for combined load effects. Although the Ferry-Borges model is still a simplified representation of the actual loading phenomenon, this model is more accurate than Turkstra's rule because it takes into consideration the rate of occurrence of the loads and their time duration. The Ferry Borges–Castanheta model assumes that the loads are constant within each time interval and are independent. However, in many practical cases, even when the intensities of the extreme load events are independent, the random effects of these loads on the structure are not independent. For example, although the wind velocities from different windstorms may be considered independent, the maximum moments produced in the piers of bridges as a result of these winds will be functions of modeling variables such as pressure coefficients as well as other statistical uncertainties that are correlated or not independent from storm to storm. In this case, Equation 1.7 has to be modified to account for the correlation between the intensities of all m possible occurrences. This can be achieved by using conditional probability functions; that is, Equation 1.7 can be used with pre-set values of the modeling factor that are assumed to be constant and then by performing a convolution over these correlated variables.

1.3.3 Wen's Load Coincidence Method

Wen's load coincidence method (Wen, 1977, 1981) is another method to calculate the probability of failure of a structure subjected to combined loads. The load coincidence method is more complicated than the two previously described approaches, but it can be used for both linear and nonlinear combinations of processes, including possible dynamic fluctuations. The load coincidence method was found to give very good estimates of the probability of failure when compared with results of simulations. Unlike the two previously listed methods, which assume independence between two different load types, the load coincidence method accounts for the rate of occurrence of each load event and the rate of simultaneous occurrences of a combination of two or more correlated loads.

The Wen load coincidence method can be represented by the following equation (Wen, 1981):

$$P(E,T) \approx 1 - \left\{ \exp\left\{ -\left[\sum_{i=1}^{n} \lambda_i p_i + \sum_{i-1}^{n-1} \sum_{j=i+1}^{n} \lambda_{ij} p_{ij} + \cdots \right] T \right\} \right\} \quad (1.8)$$

where

$P(E,T)$ = the probability of reaching limit state E (e.g., probability of failure or the probability of exceeding a response level denoted by E) in a time period T;

n = the total number of load types each designated by the subscripts i and j;

λ_i = the rate of occurrence of load type i;

p_i = the probability of failure given the occurrence of load type i only;

λ_{ij} = the rate of occurrence of load types i and j simultaneously; and

p_{ij} = the probability of failure given the occurrence of load types i and j simultaneously.

The process can be extended for three or more loads. For example, with the combination of two load types such as the combination of wind load and live load, n would be 2 and load type i may represent the live load, and load type j may represent the wind load. The rate of simultaneous occurrences of live load and wind load, λ_{ij}, would be calculated from the rate of occurrence of the wind, the rate of occurrence of the live load, and the time duration of each of the loads. The probability of failure of the bridge given that a wind and a live load occurred simultaneously is p_{ij}.

Wen's method is valid when the load intensities are pulse-like functions of time that last for very short duration. Wen's method is an extension of the one-load approach that assumes that failure events are independent and occur following a Poisson process. The loads are assumed to have low rates of occurrences, and failure events are statistically independent. Correlation between the arrival of two different load types is considered through the λ_{ij} term, while correlation between load intensities is considered through the proper calculation of the p_{ij} term. As with the case of the Ferry-Borges model, Wen's method does not account for the fact that there is cor-

relation among the probability of failure of different events that may exist because of the presence of common modeling variables. Adjustments to consider this effect are more difficult to incorporate because of the Poissonian assumptions.

This study uses the Ferry-Borges model because it provides a more intuitive approach to the load combination problem than does the mathematical formulation of Wen's load coincidence method. The Ferry-Borges model is directly implementable in Level II reliability programs as demonstrated by Turkstra and Madsen (1980) and can be modified to account for the correlation from the modeling uncertainties using conditional probability distribution functions. Similarly, Monte Carlo simulations can be easily applied to use the Ferry-Borges model, including the consideration of correlation of load effects from different time intervals.

1.4 RELIABILITY-BASED CALIBRATION OF LOAD FACTORS

The calibration process performed for the Strength I Limit State of the AASHTO LRFD (see Equation 1.1) followed traditional methods available in the reliability literature. These methods are similar to those used during the development of AISC's *Manual of Steel Construction* (1994), ACI's *Building Code Requirements for Structural Concrete ACI 318-95* (1995), and many other recently developed structural design and evaluation codes. The purpose of the theory of structural reliability is to provide a rational method to account for statistical uncertainties in estimating the capacity of structural members, the effects of the applied loads on a structural system and the random nature of the applied loads. Since absolute safety is impossible to achieve, the objective of a reliability-based calibration is to develop criteria for designing buildings and bridges that provide acceptable levels of safety.

The theory of structural reliability is based on a mathematical formulation of the probability of failure. On the other hand, the absence in the reliability formulation of many potential risks such as human errors, major defects, deliberate overloads, and so forth implies that the calculated values of risk are only notional measures rather than actuarial values. In addition, the calibration process often uses incomplete statistical information on the loads and resistance of structural systems. This is due to the limited samples of data normally available for structural applications and because each particular structure will be subjected during its service life to a unique and evolving set of environmental and loading conditions, which are difficult to estimate a priori. As an example, for bridges subjected to vehicular loads, such unique conditions may include the effect of the environment and maintenance schedules on the degradation of the structural materials affecting the strength and the particular site-dependent truck weights and traffic conditions that affect the maximum live load. The truck weights and traffic conditions are related to the economic function of the roadway, present and future weight limits imposed in the jurisdiction, the level of

14

enforcement of such limits, the truck traffic pattern, and the geometric conditions including the grade of the highway at the bridge site as well as seasonal variations related to economic activity and weather patterns. Such parameters are clearly very difficult to evaluate, indicating that the probability of failure estimates obtained from traditional reliability analyses are only conditional on many of these parameters that are difficult to quantify even in a statistical sense.

Because the reliability-based calibration gives only a notional measure of risk, new codes are normally calibrated to provide overall levels of safety similar to those of "satisfactory" existing structures. For example, during the development of bridge codes, the specification writers would assemble a set of typical member designs that, according to bridge engineering experts, provide an acceptable level of safety. Then, using available statistical data on member strengths and loads, a measure of the reliability of these typical bridges is obtained. In general, the reliability index, β, is the most commonly used measure of structural safety. The reliability index, β, is related to the probability of failure, P_f, as shown in the following equation:

$$P_f = \Phi(-\beta) \tag{1.9}$$

where Φ is the cumulative standard normal distribution function.

During the calibration of a new design code, the average reliability index from typical "safe" designs is used as the target reliability value for the new code. That is, a set of load and resistance factors as well as the nominal loads (or return periods for the design loads) are chosen for the new code such that bridge members designed with these factors will provide reliability index values equal to the target value as closely as possible.

Moses and Ghosn (1985) found that the load and resistance factors obtained following a calibration based on "safe designs" are insensitive to errors in the statistical database as long as the same statistical data and criteria are used to find the target reliability index and to calculate the load and resistance factors for the new code. Thus, a change in the load and resistance statistical properties (e.g., in the coefficients of variation) would affect the computed β values. However, the change will also affect the β values for all the bridges in the sample population of "typical safe designs" and, consequently, the average β (which is also the target β). Assuming that the performance history of these bridges is satisfactory, then the target reliability index would be changed to the new "average," and the final calibrated load and resistance factors would remain relatively the same.

The calibration process described above does not contain any preassigned numerical values for the target reliability index. This approach, which has traditionally been used in the calibration of LRFD criteria (e.g., AISC, AASHTO), has led code writers to choose different target reliabilities for different types of structural elements or for different types of loading conditions. For example, in the AISC LRFD, a target β equal to 3.5 may have been chosen for the reliability of beams in bending under the effect of dead and live loads. On the other hand, a target β equal to 4.0 may be chosen for the connections of steel frames under dead and live loads, and a target β equal to 2.5 may be chosen for beams under earthquake loads. Such differences in the target reliability index clearly reflect the economic costs associated with the selection of βs for different elements and for different load conditions, as well as different interpretations of modeling variables.

As mentioned earlier, the failure of one structural component will not necessarily lead to the collapse of the structural system. Therefore, in recent years there has been increased interest in taking into consideration the safety of the system while designing new bridges or evaluating the safety of existing ones. The same approach followed during the calibration of new codes to satisfy reliability criteria for individual structural members can also be used for the development of codes that take into consideration the system effects. For example, Ghosn and Moses (1998) and Liu et al. (2001) proposed a set of system factors that account for the system safety and redundancy of typical configurations of bridge superstructures and bridge substructures. These system factors were calibrated to satisfy the same "system" reliability levels as those of existing "satisfactory" designs.

However, whether using system reliability or member reliability indexes for criteria, the discrepancies between the observed βs for different load types raise the following question: if the reliability index β for live load is 3.5 and for earthquake loads β is 2.5, what should be the target when combining live loads and earthquake loads? On the other hand, the discrepancies in the target reliabilities can be justified using a risk-benefit argument. For example, codes should tolerate a higher risk for the design of bridges (or structures) against a particular event if the costs associated with reducing this risk are prohibitive. This risk-benefit argument can be formalized using the expected cost of a bridge or any structure defined as

$$C_T = C_I + C_F P_f \tag{1.10}$$

where

  C_T = the expected total cost,
  C_I = the initial cost for building the structure,
  C_F = the cost of failure, and
  P_f = the probability of failure.

The initial cost of a structure increases as the safety level is increased. On the other hand, the probability of failure decreases as the safety level increases. Thus, to provide an optimum balance between risk and benefit, the target reliability index that should be used is the one that minimizes the expected total cost, C_T.

Although conceptually valid, the use of Equation 1.10 in an explicit form has not been common in practice because of the difficulties associated with estimating the cost of failure, C_F. Instead, code writers have resorted to using different target reliabilities for different types of members and loads based on calibration with previous acceptable designs. A risk-benefit approach is possible if the implicit costs of a structural failure can be extracted based on current designs. Aktas, Moses, and Ghosn (2001) have demonstrated the possibility of using the risk-benefit analysis described in Equation 1.10; however, more work is still needed in evaluating the relationship between C_I and C_F before the actual implementation of this approach during the development of design codes.

Because more research is needed before the implementation of Equation 1.10 becomes possible, this report will use a traditional method of calibrating the load factors to provide a target reliability index that will be extracted based on satisfying the same safety levels of existing "satisfactory designs." For example, because the reliability index obtained for bridges that satisfy current scour design procedures is relatively small compared with that obtained for bridges designed to satisfy the wind load requirements, the target reliability level that will be used for the combination of scour and wind will be chosen to be equal to the reliability level for bridges subjected to wind alone. This will ensure that bridges subjected to combinations of wind loads and scour will have safety levels as high as those that may be subjected to high winds. A similar approach will be used when combining scour with other extreme load events. For the load combinations involving earthquakes, the reliability index obtained from earthquakes alone will be used for target. This is because the engineering community has determined that current earthquake design procedures provide sufficient levels of safety in view of the enormous costs that would be implied with any increases in the current design procedures. By using the logic described in this paragraph, a risk-benefit analysis is implicitly used in a relatively subjective manner.

1.5 RESEARCH APPROACH

The objective of this study is to develop design procedures for the application of extreme load events and the combination of their load effects in *AASHTO LRFD Bridge Design Specifications*. The load events considered in this study will include live loads, seismic loads, wind loads, ship collision loads, and scour. The design procedures will consist of a set of load factors calibrated using reliability-based methods that are consistent with the reliability methodology of the AASHTO LRFD specifications. The purpose of the AASHTO specifications is to provide procedures to proportion bridge members such that they will provide sufficient levels of safety in their service lives for any possible load type or combination of loads. This means that the bridges are expected to satisfy an accept-able level of reliability that strikes a balance between safety and cost.

To achieve the objectives of the study, this project will first review the basic reliability methodology used during previous code calibration efforts. Models to study the reliability of bridges subjected to the effects of each load taken individually will be adopted from previous bridge code calibration efforts (for the live loads and ship collisions) and from models developed during the calibration of structural codes for buildings (for wind loads and earthquake loads). Because existing specifications for scour are not based on reliability methods, a scour reliability model will be developed for the purposes of this study.

Basic bridge configurations designed to satisfy the current AASHTO specifications will be analyzed to find the implicit reliability index values for bridge design lives of 75 years and for different limit states when these bridges are subjected to live loads, wind loads, earthquakes, vessel collisions, or scour. The limit states that will be considered include column bending, shearing failure, and axial failure of bridge columns, bearing failure of column foundations, and overtipping of single-column bents.

The reliability calculations will be based on assumed probability distributions for the random variables describing the effects of extreme load events on bridge structures. These assumed distributions will be based, where possible, on available statistical data pertaining to these random variables. To study the probability of load combinations, data on the rate of occurrence of each load event, the rate of occurrence of simultaneous load events, the magnitude of each load event, and the time duration of each load event will be required. This information will be assembled from the available reliability literature, as will be described in the next chapter. Also, available statistical data on the capacity of the bridge systems to resist the applied loads will be needed. This information is available from reports describing the calibration efforts of the AASHTO LRFD specifications and other available information on the behavior of structures and foundation systems (e.g., Nowak, 1999; Poulos and Davis, 1980).

In addition to studying the reliability of the typical bridge configurations under the effects of individual extreme events, the reliability analysis will be performed for these same bridge configurations when they are subjected to the combined effects of the extreme events under consideration. Reliability methods for combining the effects of several loads have been developed by researchers in the field of reliability, as described in Section 1.3. As explained earlier, the Ferry-Borges model for load combination will be used because it provides a simple and reasonable model that is easy to implement. Modifications on the classical Ferry-Borges model will be made to account for statistical and modeling uncertainties of time-dependent and time-independent random variables.

The reliability analysis will be performed for a number of bridge configurations and for different modes of failure.

Because the extreme load events being considered are mostly horizontal loads that primarily affect the substructure of bridges, the analysis will be performed for typical bridge bents subjected to lateral loading in the transverse direction. Emphasis will be placed on failure of the columns in bending and of the foundation system subjected to lateral loads, although other failure modes such as shearing failures or axial compression of column or foundation will be considered depending on the applied load (e.g., shearing failures are important modes for barge collisions and axial compression is important for the failure of multicolumn bents). When applicable, such as in the case of bending of multicolumn bents, system effects will be taken into consideration.

The results of the reliability analyses will subsequently be used to calibrate load factors and load combination factors that will provide bridge designs with adequate levels of safety when subjected to extreme load events. To minimize the changes to the current AASHTO specifications, the load factors will be applicable to the effects of the nominal loads corresponding to the same return periods as those currently in use. The load factors will be calibrated such that bridges subjected to a combination of events provide reliability levels similar to those of bridges with the same configurations but situated in sites where one threat is dominant. Thus, the proposed load factors will be based on previous experiences with "safe bridge structures" and will provide balanced levels of safety for each load combination. As mentioned above, the target reliability indexes for the combination of events will be selected in most cases to provide the same reliability level associated with the occurrence of the individual threat with the highest reliability index. Lower reliability index target values may be justified in the cases (such as earthquake loads) when increased reliability levels would result in unacceptable economic costs. The analysis will consider structural safety as well as foundation safety and system safety will be compared with member safety. The goal is to recommend a rational and consistent set of load factors that can be used during the routine design of highway bridges. These load factors will be presented in a specifications format that can be implemented in future versions of the AASHTO LRFD specifications.

1.6 REPORT OUTLINE

This report is divided into four chapters and nine appendixes. Chapter 1 gave a review of the problem statement and an overview of the proposed research approach. Chapter 2 describes the reliability models used in this study for the different load applications and combinations of loads. Chapter 3 provides the results of the reliability analysis of typical bridge configurations subjected to the extreme events of interest and combinations of these events. Chapter 3 also provides the results of the calibration process and determines the load factors for the combinations of extreme events. Chapter 4 gives the conclusion of this study and outlines future research needs.

Appendix A summarizes the results of this study in an AASHTO specifications format that provides the load factors for the combination of extreme events. The format is suitable for implementation in future versions of *AASHTO LRFD Bridge Design Specifications*. Appendix B details the reliability model for the analysis of scour as developed by Professor Peggy Johnson for the purposes of this study. The reliability models used for the other extreme events were extracted from the literature and are similar to those used by other researchers during the calibration of various structural design codes. Appendix C provides details of the reliability calculations for a three-span bridge used as the basis for the calibration. Appendix D describes the model used for the analysis of a long-span arch bridge subjected to scour and earthquakes. Appendix E describes the analysis of a long-span bridge over the Ohio River in Maysville for vessel collisions. Appendix F describes the earthquake analysis of the Maysville Bridge. Appendix G describes the wind analysis model for both the Interstate 40 and the Maysville Bridges. Appendix H describes the reliability analysis model used for a multispan bridge subjected to earthquakes. Appendix I performs a statistical analysis of available scour data and proposes an alternative model for the reliability analysis of bridge piers under scour. The examples solved in the appendixes serve to provide details about the models used in the body of the report and also to illustrate how big projects can be specifically addressed in detail.

Appendixes A, B, C, H, and I are published herein. All appendixes (Appendixes A through I) are contained on *CRP-CD-30*, which is included with this report.

CHAPTER 2

RELIABILITY MODELS FOR COMBINATIONS OF EXTREME EVENTS

This chapter describes the models used to perform the reliability analysis of bridges subjected to extreme load events and their combinations. The chapter is divided into six sections. Section 2.1 presents the design nominal loads as specified in the current AASHTO LRFD and reviews the current code's methods for combining the effects of the extreme load events of interest to this study. Section 2.2 gives a brief review of the concepts of structural reliability theory and its application for calibration of structural design codes. Section 2.3 describes the resistance models used for member and system capacity. Section 2.4 describes the reliability models for the pertinent individual loads. Section 2.5 describes the risk analysis model used for the combination of loads. Section 2.6 concludes the chapter.

2.1 LOADS AND RETURN PERIODS IN AASHTO LRFD

The specification of load combinations in structural design codes includes the required nominal (i.e., design) loads or return periods as well as their corresponding load factors. Figure 2.1 shows the table of load factors given in the current AASHTO LRFD (Table 3.4.1). Nominal loads are usually associated with some long-term historical event that defines the loading. For example, the bridge design specifications require the design for the 50-year wind, the 2,500-year earthquake, the 75-year maximum live load, and the 100-year flood for scour. The load factors used are normally equal to or greater than 1.0 when each load is analyzed individually. When the load factor is greater than 1.0, it means that the true return period for the design event is greater than the specified nominal return period. This section reviews the current AASHTO nominal design loads and return periods for the loads of interest to this study.

2.1.1 Dead Load

The LRFD code uses a set of dead load factors that reflect the differences in the levels of uncertainties associated with estimating the weight of cast-in-place elements (particularly wearing surfaces) to those of pre-cast elements. Following common practice, the nominal (design) dead load is the best estimate obtained from design plans for new bridges or from inspection of existing bridges.

2.1.2 Live Load

Nowak (1999) calibrated the nominal HL-93 design truck loading model to match a projected expected 75-year maximum live load effect for all span ranges. The 75-year load projection was obtained based on truck weight data collected in the province of Ontario assuming a heavy volume (about 5,000 trucks per day) of truck traffic. The HL-93 load model was developed to represent the effects of live loads on short- to medium-span bridges for two-lane traffic. Multipresence factors are used to account for the differences between the HL-93 truck load effect and the expected maximum 75-year load for bridges with one lane or three or more lanes. Although the HL-93 load effects are slightly lower than the calculated maximum 75-year load effects, it is noted that the Ontario data used to calibrate the HL-93 load model contains a large level of conservativeness because it is biased toward the heavily loaded trucks. Similarly, the number of side-by-side events used to account for multiple-occurrence probabilities is also highly conservative. Moses (2001) gives a more complete discussion of these factors.

The HL-93 loading consists of a truck with similar configuration and axle weights as the HS-20 truck used in the AASHTO standard specifications (AASHTO, 1996). The truck has axle weights of 35 kN, 145 kN, and 145 kN (8, 32, and 32 Kips) along with axle spacing of 4.3 m (14 ft) and 4.3 to 9.0 m (14 to 30 ft). In addition, the HL-93 model stipulates that a lane load of 9.3 N/mm (0.64 kip/ft) should be used. For continuous bridges, two trucks are used (one in each span) along with the lane load. However, a reduction factor equal to 0.90 is included for multispan bridges. The dynamic factor of 1.33 is applied to the truck load alone excluding the effect of the lane load.

2.1.3 Earthquake Load

The design earthquake intensities specified in the current AASHTO LRFD are based on a 475-year return period for essential bridges and 2,500-year return period for critical bridges. Maps giving the design peak ground accelerations for different regions of the United States are provided by AASHTO (Figure 3.10.2; AASHTO, 1998) based on the work of NEHRP (NEHRP, 1997). The specified load factor is 1.0. The basis of the design is the inelastic response of bridges

LOAD COMBINATION LIMIT STATE	DC DD DW EH EV ES	LL IM CE BR PL LS	WA	WS	WL	FR	TU CR SH	TG	SE	EQ	IC	CT	CV
Strength I	γ_P	1.75	1.00	–	–	1.00	0.50/ 1.20	γ_{TG}	γ_{SE}	–	–	–	–
Strength II	γ_P	1.35	1.00	–	–	1.00	0.50/ 1.20	γ_{TG}	γ_{SE}	–	–	–	–
Strength III	γ_P	–	1.00	1.40	–	1.00	0.50/ 1.20	γ_{TG}	γ_{SE}	–	–	–	–
Strength IV EH, EV, ES, DW, DC ONLY	γ_P 1.5	–	1.00	–	–	1.00	0.50/ 1.20	–	–	–	–	–	–
Strength V	γ_P	1.35	1.00	0.40	0.40	1.00	0.50/ 1.20	γ_{TG}	γ_{SE}	–	–	–	–
Extreme Event I	γ_P	γ_{EQ}	1.00	–	–	1.00	–	–	–	1.00	–	–	–
Extreme Event II	γ_P	0.50	1.00	–	–	1.00	–	–	–	–	1.00	1.00	1.00
Service I	1.00	1.00	1.00	0.30	0.30	1.00	1.00/ 1.20	γ_{TG}	γ_{SE}	–	–	–	–
Service II	1.00	1.30	1.00	–	–	1.00	1.00/ 1.20	–	–	–	–	–	–
Service III	1.00	0.80	1.00	–	–	1.00	1.00/ 1.20	γ_{TG}	γ_{SE}	–	–	–	–
Fatigue—LL, IM and CE only	–	0.75	–	–	–	–	–	–	–	–	–	–	–

Figure 2.1. AASHTO LRFD load combination and load factor table (see Appendix A for definition of table symbols).

due to earthquakes. The inelastic behavior and the ductility capacity of bridge members are reflected through the use of the response modification factor *R*. Thus, material nonlinearity and system effects are indirectly taken into consideration. Also, the effect of the soil condition on the response is accounted for by using different correction factors for different types of soil. The earthquake design spectrum is given in terms of the natural frequencies of the bridge and is based on best estimates of the weight and the bridge elastic stiffness.

NCHRP Project 12-49 has proposed a comprehensive set of LRFD design specifications for the seismic design of bridges (see *NCHRP Report 472: Comprehensive Specification for the Seismic Design of Bridges* [Applied Technology Council {ATC} and the Multidisciplinary Center for Earthquake Engineering Research {MCEER}, 2002]). The proposed specifications use the NEHRP 2,500-year return period earthquake hazard maps along with the response spectra proposed by NEHRP but remove a two-thirds reduction factor that was associated with the NEHRP spectral accelerations. The two-thirds factor had been included by NEHRP to essentially reduce the 2,500-year spectrum to an equivalent 500-year spectrum for the U.S. west-coast region.

2.1.4 Wind Load

The AASHTO LRFD wind load provisions are based on the assumption that no live loads will be present on the bridge when wind velocities exceed 90 km/h (56 mph). According to the AASHTO LRFD commentaries, this assumption is based on "practical experience." The basic wind speed used is 160 km/h (100 mph) although the option of using other values is permitted. Wind speeds may be taken from the ASCE 7-95 provisions to reflect regional and geographical effects. The ASCE 7-95 maps have a return period of 50 years. The stipulated AASHTO LRFD wind load factor of 1.4 indicates that the effective return period is higher than 50 years. A comparison between the ASCE 7-95 wind maps and the data provided by Simiu et al. (1979) and Ellingwood et al. (1980) reveal that the maps provide a conservative upper limit envelope to the actual measured maximum 50-year winds.

The AASHTO LRFD contains methods to perform the wind analysis based on boundary layer theory combined with empirical observations. For wind on live loads, the AASHTO LRFD uses a distributed force of 1.46 N/mm (0.1 kip/ft) at 1800 mm (6 ft) above the roadway applied to the tributary areas that "produce load effects of the same kind." The 1.46 N/mm corresponds to the 90 km/h (56 mph) wind speed that is the limiting speed at which live load would be present on the structure.

2.1.5 Vessel Collision

The AASHTO LRFD requirements for vessel collision are based on the probability of bridge collapse using factors related to the site, the barge size, and the geometry of the waterway. According to AASHTO LRFD, the acceptable annual frequency of collapse is 0.0001 for critical bridges and 0.001 for other bridges. AASHTO LRFD stipulates that the probability of collapse should be calculated based on the number of vessels, the probability of vessel aberrancy, the geometric probability of collision given an aberrant vessel, and the probability of bridge collapse given a collision. The AASHTO LRFD provides an empirical equation to obtain the probability of bridge collapse given that a collision has occurred. Conservative assumptions are implicitly included in many of the empirical equations used in the safety check process (e.g., estimation of barge impact force), in effect further reducing the probability of collapse. The load factor specified is set at 1.0.

2.1.6 Scour

The AASHTO LRFD specifications require that scour at bridge foundations be designed for the 100-year flood storm surge tide or for the overtopping flood of lesser recurrence interval. The corresponding 100-year design scour depth at bridge foundations is estimated following the procedure recommended by FHWA using the manual known as "HEC-18" (*Hydraulic Engineering Circular No. 18: Evaluating Scour at Bridges* [Richardson and Davis, 1995]). The foundation should then be designed while taking into consideration the design scour depth. This is achieved by, for example, placing the footings below the scour depth, ensuring that the lengths of piles and pile shafts extend beyond the scour depth, and verifying that the remaining soil depth after scour provides sufficient resistance against sliding failures and overturning.

The HEC-18 manual recognizes that the total scour at a highway crossing is composed of three components: (1) long-term aggradation and degradation, (2) contraction scour, and (3) local scour. Aggradation and degradation are long-term elevation changes in the streamed of the river or waterway caused by erosion and deposition of material. Contraction scour is due to the removal of material from the bed and the banks of a channel often caused by the bridge embankments encroaching onto the main channel. Local scour involves the removal of material from around bridge piers and abutments. It is caused by an acceleration of flow around the bridge foundation that accompanies a rise in water levels that may be due to floods and other events. Both local scour and contraction scour can be either clear water or live bed. Live-bed conditions occur when there is a transport of bed material in the approach reach. Clear-water conditions occur when there is no bed material transport. Live-bed local scour is cyclic in nature because it allows the scour hole that develops during the rising stage of the water flow to refill during the falling stage. Clear-water scour is permanent because it does not allow for a refill of the hole. The focus of this report is on local live-bed scour around bridge piers that, because of its cyclical nature, is the most unpredictable type of scour.

Summary

Each loading type considered in the AASHTO LRFD provisions is represented by a nominal load intensity (i.e., a code specified live load or a geographically varying load for a given return period) and a corresponding load factor. It is noticed that the design return periods differ considerably for each type of load event (e.g., 2,500-year for earthquakes, 75-year for live load, 50-year for wind, 100-year for scour, and 1-year for vessel collisions). The reasons for the differences in the design return periods are traditional use, providing consistency with other codes, and providing assurances to the public about the safety of bridges. One should note that these return periods are only nominal because by adjusting the load factors, one could produce a common "effective return period." In addition, several of the "event intensity maps" provided as part of the specifications are based on biased envelopes of the actual measured data, and the degree of conservatism in each load analysis model varies from load type to load type. Therefore, actual failure rates may be significantly different than those implied by the nominal return periods. One major aim of structural reliability methods is to quantify these degrees of conservatism by obtaining more realistic and objective assessments of the safety levels implied in code-specified design procedures for each load type.

The calibration of the AASHTO LRFD specifications for the Strength I Limit State was undertaken using uniform reliability criteria. The code was calibrated based on the reliability of individual girders of multigirder bridges for the combination of dead and live loads. The specifications for the other extreme events of interest to this study were based on the work performed for other structural codes. In particular, AASHTO LRFD relies on the provisions and guidelines provided by

NEHRP for earthquakes, ASCE 7-95 for wind, AASHTO's *Guide Specification and Commentary for Vessel Collision Design of Highway Bridges* for vessel collisions (AASHTO 1991), and HEC-18 by FHWA for scour. In order to perform the reliability analysis and calibrate the load factors for combinations of extreme load events, statistical information on the rate of occurrence, the time duration, and the intensity of each extreme loading event is necessary. In addition, the uncertainties associated with design parameters other than the load intensities must be included. To remain consistent with the intent of the AASHTO LRFD specifications to provide safe bridge structures over an intended design life of 75 years, the load factors should be calibrated to provide consistent reliability levels for a 75-year return period, regardless of the return period that is used to define the design loads. Section 2.2 gives a review of basic concepts of reliability theory and describes the information required to perform a reliability-based evaluation of design codes. Sections 2.3 and 2.4 summarize the information available on the resistance and load models pertinent for this study and describe the methods used by various code-writing agencies and adapted in this study to account for the uncertainties associated with modeling the effect of these loads on the safety of typical bridge systems.

2.2 BASIC CONCEPTS OF STRUCTURAL RELIABILITY

The aim of the structural reliability theory is to account for the uncertainties encountered while evaluating the safety of structural systems or during the calibration of load and resistance factors for structural design codes. The uncertainties associated with predicting the load-carrying capacity of a structure, the intensities of the loads expected to be applied, and the effects of these loads may be represented by random variables. The value that a random variable can take is described by a probability distribution function. That is, a random variable may take a specific value with a certain probability, and the ensemble of these values and their probabilities are described by the distribution function. The most important characteristics of a random variable are its mean value or average and the standard deviation that gives a measure of dispersion or a measure of the uncertainty in estimating the variable. The standard deviation of a random variable R with a mean $\bar{R}$ is defined as σ_R. A dimensionless measure of the uncertainty is the coefficient of variation (COV), which is the ratio of standard deviation divided by the mean value. For example, the COV of the random variable R is defined as V_R such that

$$V_R = \frac{\sigma_R}{\bar{R}}. \tag{2.1}$$

Codes often specify nominal values for the variables used in design equations. These nominal values are related to the means through bias values. The bias is defined as the ratio of the mean to the nominal value used in design. For example, if R is the member resistance, the mean of R, namely $\bar{R}$, can be related to the nominal or design value R_n using a bias factor such that

$$\bar{R} = b_r R_n \tag{2.2}$$

where b_r is the resistance bias and R_n is the nominal value as specified by the design code. For example, A36 steel has a nominal design yield stress of 36 ksi (248,220 kPa), but coupon tests show an actual average value close to 40 ksi (275,800 kPa). Hence, the bias of the yield stress is 40/36 or 1.1.

In structural reliability, safety may be described as the situation in which capacity (strength, resistance, fatigue life, etc.) exceeds demand (load, moment, stress ranges, etc.). Probability of failure (i.e., probability that capacity is less than applied load effects) may be formally calculated; however, its accuracy depends upon detailed data on the probability distributions of load and resistance variables. Because such data are often not available, approximate models are often used for calculation.

Let the reserve margin of safety of a bridge component be defined as, Z, such that

$$Z = R - S \tag{2.3}$$

where R is the resistance or member capacity and S is the total load effect. Probability of failure, P_f, is the probability that the resistance R is less than or equal to the total applied load effect S or the probability that Z is less than or equal to zero. This is symbolized by the following equation:

$$P_f = Pr\,[R \leq S] \tag{2.4}$$

where Pr is used to symbolize the term probability. If R and S follow independent normal distributions, then

$$P_f = \Phi\left(\frac{0 - \bar{Z}}{\sigma_z}\right) = \Phi\left(-\frac{\bar{R} - \bar{S}}{\sqrt{\sigma_R^2 + \sigma_S^2}}\right) \tag{2.5}$$

where

Φ = the normal probability function that gives the probability that the normalized random variable is below a given value,

$\bar{Z}$ = the mean safety margin, and

σ_z = the standard deviation of the safety margin.

Thus, Equation 2.5 gives the probability that Z is less than zero. The reliability index, β, is defined such that

$$P_f = \Phi(-\beta) \tag{2.6}$$

which for the normal distribution case gives

$$\beta = \frac{\overline{Z}}{\sigma_Z} = \frac{\overline{R} - \overline{S}}{\sqrt{\sigma_R^2 + \sigma_S^2}}. \qquad (2.7)$$

Thus, the reliability index, β, which is often used as a measure of structural safety, gives in this instance the number of standard deviations that the mean margin of safety falls on the safe side.

The reliability index, β, defined in Equations 2.6 and 2.7, provides an exact evaluation of risk (failure probability) if R and S follow normal distributions. Although β was originally developed for normal distributions, similar calculations can be made if R and S are lognormally distributed (i.e., when the logarithms of the basic variables follow normal distributions). Also, "Level II" methods have been developed to obtain the reliability index for the cases in which the basic variables are not normal. Level II methods, often referred to as FORM or first order second moment (FOSM) involve an iterative calculation to obtain an estimate to the failure probability. This is accomplished by approximating the failure equation (i.e., when $Z = 0$) by a tangent multidimensional plane at the point on the failure surface closest to the mean value. More advanced techniques including second order reliability methods (SORMs) have also been developed. On the other hand, Monte Carlo simulations can be used to provide estimates of the probability of failure. Monte Carlo simulations are suitable for any random variable distribution type and failure equation. In essence, a Monte Carlo simulation creates a large number of "experiments" through the random generation of sets of resistance and load variables. Estimates of the probability of failure are obtained by comparing the number of experiments that produce failure with the total number of generated experiments. Given values of the probability of failure, P_f, the reliability index, β, is calculated from Equation 2.6 and is used as a measure of structural safety even for non-normal distributions. More detailed explanations of the principles discussed in this section can be found in published texts on structural reliability (e.g., Thoft-Christensen and Baker, 1982; Nowak and Collins, 2000; Melchers, 1999).

The reliability index has been used by many code-writing groups throughout the world to express structural risk. β in the range of 2 to 4 is usually specified for different structural applications (e.g., $\beta = 3.5$ was used for the calibration of the Strength I Limit State in AASHTO LRFD specifications). These values usually correspond to the failure of a single component. If there is adequate redundancy, overall system reliability indexes will be higher.

2.2.1 Code Calibration

Generally speaking, β is not used in practice for making decisions regarding the safety of a design or existing structure; rather, it is used by code-writing groups for recommending appropriate load and resistance safety factors for new structural design or evaluation specifications. One commonly used calibration approach is based on the principle that each type of structure should have uniform or consistent reliability levels over the full range of applications. For example, load and resistance factors should be chosen to produce similar β values for bridges of different span lengths, of differing numbers of lanes, of simple or continuous spans, of roadway categories, and so forth. Thus, a single target β must be achieved for all applications. On the other hand, some engineers and researchers are suggesting that higher values of β should be used for more important structures such as bridges with longer spans, bridges that carry more traffic, or bridges that, according to AASHTO, are classified as critical for "social/survival or security/defense requirements." Because higher β levels would require higher construction costs, the justification should be based on a cost-benefit analysis whereby target β values are chosen to provide a balance between cost and risk (Aktas, Moses, and Ghosn, 2001). This latter approach is still under development in order to establish the proper criteria and methods for estimating appropriate cost functions.

In many code calibration efforts, appropriate target β values are deduced based on the performance of existing designs. That is, if the safety performance of bridges designed according to current standards has generally been found satisfactory, then the reliability index obtained from current designs is used as the target that any new design should satisfy. The aim of the calibration procedure is to minimize designs that deviate from the target reliability index. Such calibration with past performance also helps to minimize any inadequacies in the database, a minimization that has been previously reported by Moses and Ghosn (1985). It was found that the load and resistance factors obtained following a calibration based on "safe existing designs" are relatively insensitive to errors in the statistical data base as long as the same statistical data and criteria are used to find the target reliability index and to calculate the load and resistance factors for the new code. In fact, a change in the load and resistance statistical properties (e.g., in the coefficients of variation) would affect the computed β values for all the bridges in the selected sample population of existing bridges and, consequently, their average β value. Assuming that the performance history of these bridges is satisfactory, then the target reliability index would be changed to the new "average," and the calibrated load and resistance factors that would be used for new designs would remain approximately the same.

The calibration effort is usually executed by code groups as follows:

- Reliability indexes are calculated for a range of bridge configurations that satisfy current code design criteria deemed to produce satisfactory performance. The calculation is based on statistical information about the randomness of the strength of members, the statistics of load intensities, and their effects on the structures.

- In general, there will be considerable scatter in such computed reliability indexes. If the existing code is believed to provide an average satisfactory performance, a target β can then be directly extracted. This is done by examining the performance and experience of selected bridge examples and averaging the β values.

- For the development of new design codes, load and resistance factors and nominal design loads (or return periods) are selected by trial and error to satisfy the target β as closely as possible for the whole range of applications.

To execute the calculation of the reliability index, one needs to obtain the statistical data for all the random variables that affect the safety margin Z of Equation 2.3, including all the uncertainties in estimating the variables that describe the member resistances and the load effects. Experimental and simulation studies have developed statistical estimates of member resistances for different types of bridge structural members. These models have accounted for the variability and uncertainties in estimating the material properties, modeling errors, differences between predicted member capacities and measured capacities, human error, and construction control. For example, Nowak (1999) followed the approach of Ellingwood et al. (1980) and represented a bridge member resistance capacity by a variable R that is the product of several variables, such that

$$R = M F P R_n \tag{2.8}$$

where

M = material factor representing properties such as strength, modulus of elasticity, capacity to resist cracking, and chemical composition;

F = fabrication error including geometry, dimensions, and section modulus;

P = analysis factor such as approximate models for estimating member capacity, idealized stress and strain distribution models; and

R_n = predicted member capacity using code-specified methods.

Equation 2.8 can be used to find the mean of R using Equation 2.2 if the total resistance bias, b_r, is set to be equal to the product of the mean values of M, F, and P.

In addition it is possible to add a system factor, λ_{sys}, that represents the capacity of the "system" to continue to carry loads after the failure of the first member. Thus, the system capacity, R_{sys}, may be represented as follows:

$$R_{sys} = \lambda_{sys} M F P R_n. \tag{2.9}$$

The resistance model of Equations 2.8 and 2.9 does not directly account for member deterioration or other changes with time. Thus, all the variables are time-independent random variables.

For a bridge member (or structural system) to be safe, the resistance should be large enough to withstand the maximum load effect that could occur within the structure's service life. Estimating the effects of the maximum loads involves a number of random variables, which may often be associated with large levels of modeling uncertainties. In particular, the intensities of the maximum loads are time-dependent random variables in the sense that longer service lives imply higher chances that the structure will be subjected to a given extreme load level. On the other hand, the projection of limited load intensity data, collected from previous measurements over short periods of time, to future return periods is associated with various levels of statistical modeling uncertainties. In addition, modeling the structure's response to the applied loads and estimating the variables that control the effects of the loads on the structure are associated with high levels of uncertainty that are independent of the return period. These modeling uncertainties are often represented by time-independent random variables. Thus, the effect of a particular load type, i, on a structural member may be represented by an equation of the form

$$S_i = \lambda_i f_i (\lambda_{Qi} C_{ji} Q_i) \tag{2.10}$$

where

S_i = the load effect for load type i;

λ_i = the analysis modeling factor that accounts for differences between measured load effects and predicted load effects;

$f_i( )$ = the analysis prediction model that converts load intensities into load effects;

Q_i = the projected intensity variable of load type i for the return period of interest;

λ_{Qi} = the statistical modeling variable that accounts for the limitations in predicting the value of Q_i; and

C_{ji} = the analysis variables such as bridge material and geometrical properties required for executing the analysis for load type i.

Several such variables, each represented by the index, j, may be required to execute the load effect analysis. As mentioned above, all the variables in Equation 2.10 may be considered random where Q_i is a time-dependent random variable and the remaining variables are time-invariant.

The probability density of the load intensity, Q_i, for a given return period, t, can be calculated by studying the probability that Q_i will exceed a given value within t. Assuming that the occurrence of load events follows a Poisson model, the probability that the load intensity will exceed a value x, within a period, t, is represented by $(1 - F_{Qi,t}[x])$, which may be approximated as

$$\Pr(Q_i > x; \; T < t) = 1 - F_{Qi,t}(x) = 1 - e^{(-tp)} \tag{2.11}$$

where p is the rate of exceedance per unit time. p is equal to the probability of exceeding x when t equals 1.0:

$$p = \Pr(Q_i > x) = 1 - F_{Qi}(x). \qquad (2.12)$$

For extreme values of x when the values of $F_{Qi}(x)$ are close to 1.0, and when p is calculated for one unit of time while the return period, t, consists of m units of time, Equation 2.11 can be approximated as

$$\Pr(Q_i > x;\ T < t) = 1 - F_{Qi,t}(x) = 1 - e^{(-tp)} \approx 1 - (1-p)^m$$
$$= 1 - (F_{Qi}(x))^m \qquad (2.13)$$

or

$$\Pr(Q_i < x;\ T < t) = F_{Qi,t}(x) \approx F_{Qi}(x)^m \qquad (2.14)$$

In other words, Equation 2.14 indicates that the cumulative probability function for a return period of time, t, may be approximated by raising the cumulative probability function of the basic time period to the power, m. For example, assuming that $F_{Qi}(x)$ is the probability distribution of Q_i for 1 day, the probability distribution for a return period of 100 days can be calculated from Equation 2.11 or Equation 2.14 by setting m equal to 100. On the other hand, if $F_{Qi}(x)$ is the probability distribution for one event, then assuming n events per day, the probability distribution of the load intensity for a 100-day return period is calculated by setting m equal to $100n$.

Equations 2.11 through 2.14 are valid for studying the effect of one individual load of type i. However, the probability that two or more load events will occur simultaneously within the service life of the structure is a function of the rate of occurrence of each event, the time duration of the events, and the correlation between the events. Even when two load types can occur simultaneously, there is little chance that the intensities of both events will simultaneously be close to their maximum lifetime values. Therefore, the reliability calculations must account for all the random variables associated with estimating the maximum values of each load type and the possibility of load combinations. Available methods for studying the load combination problem have been discussed in Chapter 1. In particular, the Ferry-Borges model that is used in this study is described in detail in Section 2.5.

In order to execute the reliability calculations based on the Ferry-Borges model, statistical data on structural member and structural system resistances as well as data on the load intensities, their rate of occurrence, and their time duration are required. In addition, the uncertainty in modeling the effect of the loads on the structures should be considered. The next two sections give summaries of the statistical models used in this study based on the work available in the reliability literature. The models used are adopted from previous studies that led to the development of bridge design specifications or other structural design codes.

2.3 RESISTANCE MODELS

The object of this project is to study the reliability of highway bridges subjected to extreme load events and their combinations. In addition to live loads, the extreme events under consideration consist of wind loads, earthquake loads, ship collisions, and scour. Except for live loads and scour, all these events primarily produce horizontal loads on bridges. The bridge substructure consisting of the columns in a bridge bent and the foundation is the primary system that resists the applications of horizontal loads. The occurrence of scour, which is the erosion of the soil material around bridge piers, would also jeopardize the safety of the foundation system. Therefore, the focus of this study will be on the capacity of bridge substructures to resist the effect of lateral forces induced by the extreme events under consideration.

The intensities of the extreme load events are time-dependent random variables in the sense that the longer the bridge structure is operational, the higher the chances are that an extreme event with a high level of intensity will occur, causing the failure of the structure. In addition, as mentioned above, modeling the effect of these load events and the resistance capacity of the structure is associated with varying degrees of uncertainties because of the limitation of the analysis models that are commonly used as well as the difficulty of determining the exact properties of the structure and other factors that affect its behavior. The approach used in this study to account for these modeling uncertainties and material variability is by assuming that many of the factors that control the behavior and the response of bridges can be treated as random variables. The statistical uncertainties and the biases associated with estimating these variables are normally obtained from comparisons of analytical models with test results and by studying the data assembled from various measurements of material properties and bridge member and system responses. The object of this current project is not to collect such statistical data but to use the "most widely accepted" models for the needs of this study. A model is generally judged to be "acceptable" if it has been applied during recent code calibration efforts. This section provides a review of the structural and foundation resistance models that are used during recent studies dealing with the calibration of design codes or that are most commonly used in practice. These models will be adapted for use in this project to calibrate the load factors for the design of highway bridges and to study the reliability of such bridges when they are subjected to extreme events and combinations of extreme events.

2.3.1 Bridge Column Resistance Capacity

The concrete columns of a bridge bent are subjected to lateral loads from the extreme events as well as from vertical gravity loads caused by the permanent weight and live load.

The safety of a bridge column under combined axial loads and moment-causing lateral loads is defined by a *P-M* interaction curve (axial load versus moment curve). Figure 2.2 shows a typical *P-M* curve for a concrete column. In bridge engineering practice, most columns are designed such that the applied combined loading remains in the vicinity of the balanced point, that is, in the region in which the effect of the axial load does not significantly reduce the moment capacity of the column. For this reason, it is herein assumed that when the substructure is subjected to lateral loads, column failures will primarily be due to bending. Hence, the statistical data provided by Nowak (1999) for calibrating AASHTO LRFD resistance factors for bending of concrete members will be used when checking the safety of bridge columns subjected to lateral loads. When only vertical loads are applied, the dominant loading is axial compression. In this case, the model proposed by Ellingwood et al. (1980) will be used.

During the AASHTO LRFD calibration, Nowak (1999) used a bias of 1.14 and a COV of 13% for the moment capacity of reinforced concrete members. A "bias" is defined as the ratio of the mean value to the nominal or design value. This means that the average moment capacity of a bridge column will generally be different than the value calculated using the procedures outlined by the AASHTO LRFD code. There will also be a scatter in the moment capacities of columns designed to the same specifications and constructed following the same common construction procedures. The average value of these capacities will be 1.14 times the specified nominal capacity. "COV" is defined as the standard deviation divided by the mean value. Using Equation 2.1, a COV of 13% means that

the standard deviation will be equal to 0.148 (0.13 × 1.14) times the specified value. The biases and COVs, used during the AASHTO LRFD calibration, account for uncertainties in the material strength, geometric dimensions, and analysis. During the calibration process, Nowak (1999) used a lognormal probability distribution to model the member resistance and to describe the scatter of the bending moment capacities.

For axial loading of concrete columns, Ellingwood et al. (1980) used a lognormal distribution with a bias of 1.05 and a COV of 16%. When bridge columns are subjected to concentrated transverse forces such as when a ship or a barge collides into a column, the column may fail in shear. For shear failures of concrete columns Nowak (1999) uses a bias of 1.40 and a COV of 17% for columns with no steel shear reinforcement. When the columns are reinforced with steel, a bias of 1.2 and a COV of 15.5% are used.

The bending failure of one column in a multicolumn bent does not necessarily lead to the collapse of the substructure. In fact, Liu et al. (2001) have shown that multicolumn bents formed by columns confined with lateral reinforcing steel will, on the average, be able to withstand 30% additional load beyond the load that causes the first column to reach its ultimate bending capacity. The COV was found to be on the order of 12.5%. Substructure systems composed of unconfined multiple columns will on the average provide 15% more bending capacity with a COV of 8%. The distribution of the system's modeling factor is assumed to be normal. This issue is further discussed in Section 2.3.3. The characteristics of the random variables describing the capacity of the columns of bridge substructures are summarized in Table 2.1.

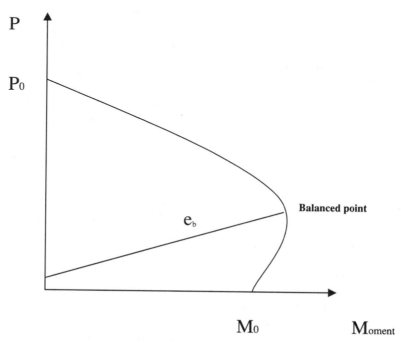

Figure 2.2. Strength interaction diagram for concrete columns.

TABLE 2.1   Input data for concrete column capacity

Variable	Bias	COV	Distribution Type	Reference
Moment capacity, M_{col}	1.14	13%	Lognormal	Nowak (1999)
Axial capacity, P_{col}	1.05	16%	Lognormal	Ellingwood et al. (1980)
Shear capacity no steel, V_{col1}	1.40	17%	Lognormal	Nowak (1999)
Shear capacity with steel, V_{col2}	1.20	15.5%	Lognormal	Nowak (1999)
System capacity for bending of unconfined multicolumn bents, $\lambda_{sys,u}$	1.15	8%	Normal	Liu et al. (2001)
System capacity for bending of confined multicolumn bents, $\lambda_{sys,c}$	1.30	12.5%	Normal	Liu et al. (2001)

2.3.2 Foundation/Soil Resistance Capacity

Another important failure mode that may jeopardize the safety of bridge substructures is the failure of the foundation system. This failure could be due to the failure of the foundation structural elements (e.g., piles) or failure in the soil. The capacity of the soil to resist failure is controlled by the specific weight of the soil; the type of soil material (e.g., clay or sand) and the depth; and the size and type of the foundation (e.g., footings on piles or pile shaft).

For most of the examples treated in this study, the foundation system is considered to be a drilled shaft pile in sands. This assumption is made in order to keep the analysis of the foundation reasonably simple and consistent with the overall objective of this study. The response of the column to lateral loads can then be modeled as shown in Figure 2.3. The simplified model used assumes an ultimate soil resistance capacity related to Rankine's passive pressure as described by Poulos and Davis (1980).

In Figure 2.3, F is the applied lateral force, e gives the height of the column above ground level, L is the foundation depth, and P_p is the passive resultant resisting force of the soil (produced by the triangular soil pressure resisting the motion). For long piles, the maximum bending moment in the pile shaft occurs at a distance f below the soil surface where the force from the soil pressure, up to depth f, is equal to the applied lateral force F. If f is larger than the foundation depth, L, the pile acts as a short pile, the controlling mode of failure is rigid overtipping about the base of the pile, and the moment arm is $H = L + e$. This is shown in Figure 2.3(a). If f is smaller than L, then the maximum moment will occur at a depth f, and the controlling mode of failure is bending of the pile. In this case, the moment arm is $H = f + e$ and the pile acts as if

it is fixed at a distance f below the soil level. This is shown in Figure 2.3(b).

According to Poulos and Davis (1980), the nominal active force P_p for sandy soils is given as

$$P_p = \frac{3\gamma D K_p L^2}{2} \tag{2.15}$$

where

γ = the specific weight of sand,
L = the depth of the pile,
D = the diameter of the pile, and
K_p = the Rankine coefficient, which is given by

$$K_p = \frac{1 + \sin(\phi_s)}{1 - \sin(\phi_s)} \tag{2.16}$$

where ϕ_s is the angle of friction for sand.

When the depth of the soil is such that $f < L$, the pile system is simply an extension of the bridge column and can be treated in the same manner as the column, using the same bias and COV given by Nowak (1999) for bridge members in bending.

The dominant factors that control the safety of the soil are the specific weight of the soil, the angle of friction, and the modeling uncertainties associated with using the Rankine model. Poulos and Davis (1980) and Becker (1996) mention that for the lateral soil pressure on pile shafts, the Rankine model is conservative, providing a bias of about 1.5 when compared with measured data with a COV on the order of 20%. Becker (1996) also mentions that the specific weights

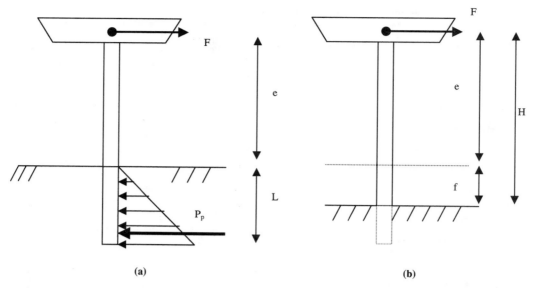

Figure 2.3.   Free body diagram of bridge foundation system: (a) free body diagram for a short pile, dominant failure mode is tipping at base; (b) diagram for long pile, dominant failure mode is bending at distance f below soil level.

of soils will have variability with a COV equal to 7% compared with the measured value while estimates of angle of friction are associated with a COV of 13%. For the vertical bearing capacity, Poulos and Davis (1980) indicate that the applicable pile design formulas provide a bias of 1.0 when compared with pile tests with a COV on the order of 25%. Because most of the loads under consideration are dynamic in nature, it is proposed to follow the approach of Bea (1983) and apply a cyclic bias of 1.0 with a COV of 15% when analyzing the resistance of soils for foundations subjected to repeated cyclic forces. The primary random variables that control the estimation of the lateral soil resistance of the pile-

shaft system under lateral loads are summarized in Table 2.2. A normal distribution is assumed for all the random variables.

2.3.3 System Capacity

The analysis of multiple-column bents produces different moments in each column because of the effect of the dead load and the presence of axial forces. An extensive analysis of different bent configurations founded on different soil types was performed for NCHRP Project 12-47 (Liu et al., 2001). The results showed that, because of the load redistri-

TABLE 2.2   Input data for soil-related random variables

Variable	Bias	COV	Distribution Type	Reference
Unit weight of soil, γ	1.0	7%	Normal	Becker (1996)
Angle of friction, ϕ_s	1.0	13%	Normal	Becker (1996)
Rankine earth pressure model for lateral force on piles, K_p	1.5	20%	Normal	Poulos and Davis (1980) Becker (1996)
Soil bearing capacity, P_{soil}	1.0	25%	Normal	Poulos and Davis (1980)
Cyclic effects, λ_{cyc}	1.0	15%	Normal	Bea (1983)

bution and the presence of ductility, multiple-column bents with confined concrete would on the average fail at loads up to 30% higher than the loads that make the first column reach its ultimate member capacity with a COV of about 12.5%. Systems with unconfined columns give system reserve ratios of about 1.15 (15% additional system capacity compared with first member capacity) with a COV of about 8%.

It is noted that system reserve ratio should be included to study the safety of the system (as compared with member safety) only for the cases in which a linear elastic analysis is performed. In the cases in which a full plastic analysis is performed, the system reserve is automatically taken into consideration.

2.4 LOAD MODELS

As mentioned earlier, the intensity of the extreme load events are time-dependent random variables in the sense that the longer the bridge structure is operational, the higher the chances are that an extreme event with a high level of intensity will occur causing the failure of the structure. However, modeling the effects of these loads is also associated with varying levels of uncertainties because of the limited available data, the limitations of the commonly used analysis models, and the difficulty of estimating the exact properties of the structure and other factors that affect its behavior. The approach followed in this project is to account for the analysis model uncertainties by assuming that many of the parameters that control the behavior and the response of bridges can be treated as random variables. The biases and COVs associated with estimating these variables are normally obtained from comparisons of analytical models to test results and by studying the data assembled from various measurements of bridge member and system responses. The object of this current project is not to collect such statistical data but to use "widely accepted" statistical models that have been applied during recent code calibration efforts for the needs of this study. This section provides a review of the models used in this study to calibrate the load factors for the design of highway bridges subjected to the combination of extreme events.

2.4.1 Gravity Loads

Dead Loads

The dead load biases and COVs used by Nowak (1999) during the reliability calibration of the AASHTO LRFD are

1. A bias of 1.03 with a COV of 8% for factory-made members,
2. A bias of 1.05 with a COV of 10% for cast-in-place members, and
3. For the wearing surface, a 88.9-mm (3.5-in.) average asphalt thickness along with a COV of 25%.

Nowak (1999) also assumes that the total load effect (live load + dead load) follows a normal distribution. Ellingwood et al. (1980) suggest that the dead load follows a normal distribution. In the calculations performed in this study, a common bias of 1.05 and a COV of 10% are used for the dead load using a normal distribution.

Live Loads

A major portion of the AASHTO LRFD calibration effort was spent on studying the effect of live loads. The statistical database for live loads was primarily obtained from truck weight histograms collected in Ontario (Nowak and Hong, 1991). Also, a limited set of weigh-in-motion data collected in Michigan (Nowak and Hong, 1991) and a weigh-in-motion study supported by FHWA (Goble et al., 1991) were available. Using the Ontario statistical database, Nowak (1999) calculated the target reliability index for member capacity, β_{target}, to be 3.5. This value is the average reliability index obtained for a representative sample of components of different bridge configurations designed to satisfy the AASHTO standard specifications (1996) with the HS-20 design load. Load and resistance factors were calibrated for the AASHTO LRFD using the same statistical database and using the 3.5 target value. The assumption of the code writers was that HS-20 bridges have performed satisfactorily under current live loading conditions and, thus, the new code should be calibrated to provide on the average a similar level of safety.

Only "heavy" trucks were measured in the Ontario survey. Nowak (1999) does not provide the actual truck weight histogram, but Moses (2001) found that the weights of the "heavy trucks" used in the AASHTO calibration effort approach a normal distribution with a mean of 300 kN (68 kips) and a standard deviation of 80 kN (18 kips) (COV = 26.5%). Nowak (1999) indicates that a typical site will have an average daily "heavy" truck traffic of about 1,000 trucks per day. Nowak (1999) also assumes that about 1 in every 15 "heavy" trucks will be side-by-side with another "heavy" truck. This produces a rate of side-by-side "heavy" truck occurrences on the order of 6.67%. On the other hand, data collected by Moses and Ghosn (1985) have shown that the truck population (for all trucks) on Interstate highways is on the order of 2,000 to 3,000 trucks per day (it may even reach 5,000 trucks per day in some cases) but that the percentage of side-by-side events for all the truck population is on the order of 1% to 3%. This percentage, however, may depend on site conditions and truck traffic volume. This comparison would indicate that Nowak's data are highly biased toward the conservative side.

The statistics on truck weights (mean = 300 kN and standard deviation = 80 kN) and truck daily rates (1,000 trucks per day) are used to find the expected maximum truck load effect for different return periods using Equations 2.11 and 2.14. The calculations show that these statistics would produce results that replicate the results provided by Nowak

(1999) for the maximum extreme load effects for various return periods. Nowak (1999) does not provide the COVs for the different return periods that he considered, but the calculations using Equations 2.11 and 2.14 show that the COV for the truck load intensity decreases as the return period increases. Calculations for a 75-year return period using the mean truck weight of 300 kN and a standard deviation of 80 kN show that the 75-year maximum truck weight intensity that a bridge will be subjected to will be 734 kN (165 kips) with a COV less than 3%.

The 3% COV calculated above reflects the fact that for very long projection periods and for large number of truck occurrences, the chances that the bridge will be subjected to a high truck load become more certain assuming a high level of confidence in the input data base. On the other hand, because of statistical and modeling uncertainties, the final COV for the maximum combined load effects should be higher than 3%. For example, assuming two lanes of traffic, Nowak (1999) used a final COV for the truck load effects on the order of 19% to 20%. The final COV values used by Nowak (1999) account for the uncertainties in estimating the maximum live load, the load distribution to individual members, and the dynamic amplification factor. It is herein believed that this 19% associated with a bias of 1.0 must also account for the site-to-site variability (see Moses and Ghosn, 1985).

Although the final COV of 19% to 20% used by Nowak (1999) accounts for the uncertainty in estimating the dynamic amplification effects of moving vehicles on a flexible bridge superstructure, Nowak also gives the statistical information on the dynamic amplification factor separately. According to Nowak (1999), the measured dynamic effect (dynamic amplification factor) has a mean value equal to 9% of the static effect for two lanes of traffic with a COV of 6%. For one lane of traffic, the dynamic amplification factor is 13% of the static loads with a COV of 10%.

Given a 75-year maximum expected truck intensity COV of 3%, a dynamic amplification COV of 10%, and a final COV of 19% to 20%, it is concluded that the modeling uncertainties including site-to-site variability are associated with a COV on the order of 18%. A lognormal distribution is used to represent the modeling uncertainties because these involve the multiplicative effects of several factors including an axle distribution factor that accounts for different truck configurations, an analysis uncertainty factor, a statistical uncertainty factor, and a site-to-site variability factor (see also Moses and Ghosn, 1985).

The HL-93 live load model of the AASHTO LRFD specifications (1998) is intended for a 75-year design life. Several tables provided by Nowak (1999) give the bias between the projected maximum 75-year effects for one lane of traffic and the HL-93 truck load effects for simple span and multispan bridges. Because this project places emphasis on the reliability analysis of bridge bents, the focus is on multispan bridges. Table B-16 of Nowak's report lists the variation of the bias as a function of the return period for the negative moment of two equal continuous spans. For the purposes of

this study, it is assumed that these biases are valid for all responses. In particular, the same biases of Table B-16 would be used for the reaction at the support. Trains of trucks rather than single occurrences control the loading of continuous span bridges. To extract the effect of one loading event, consisting of a train of trucks in one lane of traffic, the same approach proposed by Moses (2001) is followed. The purpose of the calculations is to find the bias for a single load event that would reproduce the results of Table B-16 of Nowak (1999) for one lane of traffic. Using Equations 2.11 and 2.14, it is found that the biases provided in Table B-16 of Nowak (1999) for the different return periods are exactly matched if the one load event is assumed to follow a normal distribution and is associated with a bias of 0.79 with a COV of 10%. The calculations assume that there will be 1,000 loading events each day with each loading event consisting of a train of trucks in one lane of traffic. The 0.79 bias is relative to the effect of one lane loading of the HL-93 live load model. For two lanes of traffic loaded simultaneously, the bias relative to the effect of one lane loading of the HL-93 model is found to be 1.58 with a COV of 7%. The calculations assume that there will be 67 events of two lanes loaded simultaneously. Using Equation 2.11 and 2.14, it is found that the two-lane loading will produce a 75-year maximum load effect equal to 2×0.88 (= 1.76) times the load effect of one lane of traffic. The 2×0.88 factor is similar to the factor reported by Nowak (1999), who has suggested that a factor equal to 2×0.85 (= 1.70) gives a reasonable approximation to the effect of two-lane traffic relative to one-lane traffic.

In summary, to perform the reliability calculations for the needs of this study, the following assumptions are made:

1. A "heavy" truck loading event will, on the average, consist of a truck with a weight of 300 kN (68 kips) and a standard deviation of 180 kN (8 kips) (COV = 26.5%). In a typical day, there will be 1,000 "heavy loading events" (i.e., the rate of occurrence is 1,000 events per day).
2. Two side-by-side "heavy" truck events will occur at a rate of 67 events per day (i.e., 1,000/15) and, on the average, the weights of the two trucks will be 600 kN (136 kips) with a COV of 19%.
3. Multispan bridges will be subjected to 1,000 one-lane load events per day with a mean load effect value equal to 0.79 times the effect obtained from the HL-93 loading. The COV for one-lane load effects is 10%.
4. Multilane multispan bridges will be subjected to 67 two-lane load events with a mean value equal to 1.58 times the effect obtained from one HL-93-lane load. The COV for two-lane load effects is 7%.
5. The dynamic amplification factor for one lane of traffic is 1.13 with a COV of 10%. For two lanes of traffic, the amplification factor is 1.09 with a COV of 6%.
6. A modeling variable is used with a mean value of 1.0 and a COV of 18%.
7. The truck load effects and dynamic factors are assumed to follow normal distributions as suggested by Nowak

(1999). The modeling variable is assumed to follow a lognormal distribution.

The live load model proposed in this section provides results that are consistent with the database used by Nowak (1999) when calibrating the AASHTO LRFD specifications. These are summarized in Table 2.3. It is noted that the dead load, the live load modeling factor, and the dynamic amplification factors are all time-independent random variables. Only the intensity of the applied live load will increase with the bridge's exposure period.

2.4.2 Earthquakes

The analysis of the response of a bridge structure to earthquakes involves a number of random variables. These variables are related to the expected earthquake intensity for the bridge site; the number of earthquakes at the site; the natural period of the bridge system; the spectral accelerations for the site, including soil properties; the nonlinear behavior of the bridge system; and the modeling uncertainties associated with current methods of analysis. The emphasis of the analyses performed in this project is on the horizontal motion of bridge bents caused by earthquakes. A discussion on each variable follows.

Intensity of Earthquake Accelerations

The USGS mapping project developed maps providing the peak ground accelerations (PGAs) at bedrock level for various sites throughout the United States in terms of the proba-

bility of exceedance in 50 years (Frankel et al., 1997). These maps are available for probabilities of exceedance of 2% and 10% in 50 years. The 2% probability of exceedance in 50 years corresponds to an earthquake return period of about 2,500 years. The 10% in 50 years corresponds to a return period of about 500 years. These return periods are normally used as the bases of current bridge design practice. The probabilities of exceedance in 50 years can directly be related to the maximum yearly earthquake levels using Equations 2.11 or 2.14. In addition, Frankel et al. (1997) provide a number of curves describing how the probability of exceedance in 1 year varies as a function of PGA for a number of sites. Figure 2.4 gives the yearly probability of exceedance for a range of PGAs and for a number of representative sites provided by Frankel et al. (1997) for the areas with the following zip codes: 10031 in New York City, 38101 in Memphis, 55418 in St. Paul, 98195 in Seattle, and 94117 in San Francisco.

Rate of Earthquake Occurrences

The number of earthquakes varies widely from site to site. The USGS mapping project also provides the expected number of earthquakes at particular sites. For example, the average number of earthquakes in 1 year for the five earthquake sites listed above varies from about 8 per year to 0.009 per year. For example, San Francisco may be subjected to about 8 earthquakes per year while Seattle may be subjected to 2 earthquakes. The rates for the other sites are usually less than 1.0: Memphis will witness one every 2 years (yearly rate = 0.50), New York City one every 2.5 years (yearly rate = 0.40), and St. Paul one in 111 years (yearly rate = 9×10^{-3}).

TABLE 2.3 Input data for dead and live load random variables

Variable	Bias	COV	Distribution Type	Reference
Dead load, F_{DC}	1.05	10%	Normal	Nowak (1999)
Live load modeling factor, λ_{LL}	1.0	18%	Lognormal	Moses and Ghosn (1985)
Live load multispan one event in one lane (relative to one lane of HL-93 load), $I_{LL,1}$	0.79	10%	Normal	Calculated based on Nowak (1999) and Moses (2001)
Live load one event in multilane (relative to one lane of HL-93 load), $I_{LL,2}$	1.58	7%	Normal	Calculated based on Nowak (1999) and Moses (2001)
Dynamic amplification one lane, $I_{IM,1}$	1.13 (=mean value)	10%	Normal	Nowak (1999)
Dynamic amplification two lanes, $I_{IM,2}$	1.09 (=mean value)	6%	Normal	Nowak (1999)

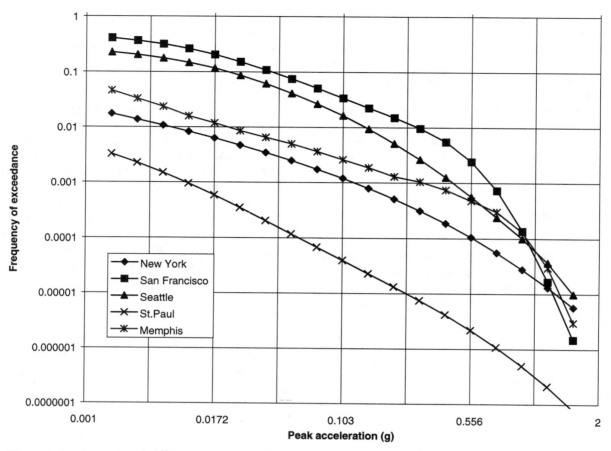

Figure 2.4. Annual probability of exceedance curves for PGA (based on USGS website).

Natural Period of Bridges

The natural period of a bridge depends on many parameters, including the type and the characteristics of the bridge structure and bridge foundation and the stiffness of the soil. Takada, Ghosn, and Shinozuka (1989) have suggested that the average value of the period is about 1.08 times the value calculated using design methods (bias = 1.08) with a COV on the order of 20%. These values account for the soil-structure interaction (SSI) and other analysis effects. The values provided by Takada, Ghosn, and Shinozuka (1989) are primarily for buildings. The Takada, Ghosn, and Shinozuka (1989) data should be applied on structural models that did not include the effects of SSIs because the 1.08 bias accounts for the effects of SSI. Because the analysis model used in this study includes the effects of SSI, a lower bias should be used (Stewart, Seed, and Fenves, 1999).

When SSI models are included in the analysis (as is the case in the models used in this project), the variation between the measured periods and the predicted periods appears to be smaller, and the bias is reduced to a value close to 1.0 (Stewart, Seed, and Fenves, 1999). This phenomenon is shown in Figure 2.5, which is adapted from the paper by Stewart, Seed, and Fenves (1999). The ordinate of the fig-

ure gives the ratio of the difference between the predicted period, $\hat{T}_{pre}$, and the measured period, $\hat{T}$, divided by $\hat{T}$, where the periods account for the flexibility of the foundation because of SSI. The abscissa is the ratio of the structure height, h, over the product of the soil shear-wave velocity, V_s, times the measured period of the fixed structure, T. The abscissa is thus the inverse of the ratio of the soil-to-structure stiffness defined as σ. The figure shows that, on the average, the predicted periods accounting for SSI are reasonably similar to the values measured in the field. The bias is found to be 0.99 and the COV on the order of 8.5%. It is noted that the comparison is made for buildings rather than for bridges and that most of these sites had stiff soils. Also, the natural period of the fixed structural system, T, used in the abscissa is inferred from field measurements.

Chopra and Goel (2000) developed formulas for estimating the natural periods of buildings based on measured data. The spread in the data shows a COV on the order of 20% for concrete buildings and slightly higher (on the order of 23%) for steel frame buildings. Haviland (1976) suggested that a bias of 0.90 and a COV of 30% are appropriate for the natural period of structural systems.

Based on the review of the above references, it is herein decided to use a bias of 0.90 and a COV of 20% for the period

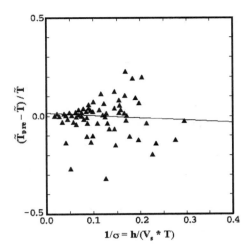

Figure 2.5. Variation of predicted natural period of structural systems including SSI compared with measured values (based on Stewart, Seed, Fenves, 1999).

of the system. The 0.90 bias, similar to that observed by Haviland (1976), is justified based on the fact that the analysis performed to calculate the period of the bridge system uses the nominal value for the modulus of elasticity of the concrete, E_c. In reality, the actual modulus of the concrete will be higher than the nominal value and, thus, the predicted stiffness larger than the actual stiffness, producing a lower actual period than predicted. A correction factor of 1.20 to 1.30 on the concrete modulus is often used in engineering practice that would justify the 0.90 bias $\left(0.90 \approx 1/\sqrt{1.25}\right)$. The COV of 20% used herein corresponds to the values observed by Chopra and Goel (2000) for buildings and the value used by Takada, Ghosn, and Shinozuka (1989). Also, the 20% COV provides an average value between the COV values from the various references studied. It may be argued that the prediction of the period for bridges may be less uncertain than that for buildings. However, the data collected by the various researchers cited in this section did not indicate any appreciable difference in the level of uncertainty because of building heights or sizes. Thus, it may be reasonable to assume that the COV on the period for bridges is also on the order of 20%. The proposed bias and COV are meant to account for both the uncertainties in the structural properties and the SSI parameters.

Mass Applied on the Column

The dead weight effect on bridge members was found by Nowak (1999) to be on the order of 1.05 times the nominal (design) weight with a COV of 10%. This COV, however, reflects the effects of the structural analysis and the uncertainty in estimating the weight. To account for the uncertainty in the weight alone, a COV of 5% is used. The probability distribution is assumed to be normal following the models used by Ellingwood et al. (1980) and Nowak (1999). The

variability in the mass and the weight considered here are used for calculating the earthquake inertial forces.

Seismic Response Coefficient

The USGS mapping project (Frankel et al. 1997) provides spectral accelerations for different sites within the United States. The spectral accelerations are given for periods of 0.2, 0.3, and 1.0 sec. Table 2.4 lists these values along with the corresponding PGAs for different sites. The sites considered are for the following zip codes: 10031 in New York City, 38101 in Memphis, 55418 in St. Paul, 98195 in Seattle, and 94117 in San Francisco. The spectral accelerations provided in Table 2.4 are for single-degree-of-freedom (SDOF) systems founded on bedrock. The values are given for only three natural periods—namely, 0.2, 0.3, and 1.0 sec—in addition to the PGA. The values shown in Table 2.4 are in % g where g is the acceleration caused by gravity. The spectral accelerations are given for a probability of exceedance of 2%, 5%, and 10% in 50 years.

NEHRP (1997) proposed a method to use the information provided in Table 2.4 to develop acceleration response spectra that are valid for systems with different natural periods. The NEHRP response spectra can be described by curves with the shape shown in Figure 2.6. In Figure 2.6, the ordinate gives the final normalized spectral acceleration, S_a (normalized with respect to g), and the abscissa is the natural period of the system, T. S_{Ds} is the maximum spectral acceleration that is the spectral acceleration for the short period $T_s = 0.2$ sec. S_{D1} is the spectral acceleration for a period $T_1 = 1$ sec. All spectral accelerations are given in function of the acceleration due to gravity, g. T_0 gives the period at which the maximum spectral acceleration is reached. T_s gives the period at which the spectral acceleration begins to decrease below the maximum value. When the period T is less than T_0, the spectral acceleration increases linearly. When the period T is greater than T_s, the spectral acceleration is inversely proportional to T. The values of S_{Ds} and S_{D1} are obtained from the spectral accelerations given in Table 2.4 adjusted to account for the site's soil properties.

The site soil coefficients are obtained from the NEHRP provisions as F_a for short periods and F_v for 1 sec from Table 4.1.2.4.a and 4.1.2.4.b of NEHRP (1997). NCHRP Project 12-49 (ATC and MCEER, 2002) is proposing to use a modification on the NEHRP spectra such that the maximum earthquake spectral response accelerations for short-period (0.2-sec) S_{DS} and for the 1-sec period S_{D1}, adjusted for the proper soil profile, are obtained from

$$S_{DS} = F_a S_S \qquad \text{and} \qquad S_{D1} = F_v S_1. \qquad (2.17)$$

Critical periods on the response spectrum (see Figure 2.6) are obtained from

$$T_0 = 0.20 S_{D1}/S_{DS} \qquad \text{and} \qquad T_s = S_{D1}/S_{DS}. \qquad (2.18)$$

TABLE 2.4   Probabilistic ground motion values, in %g, for five sites (USGS website)

	10% PE in 50 Yr	5% PE in 50 Yr	2% PE in 50 Yr
San Francisco			
PGA	52.65	65.00	76.52
0.2 sec S_a	121.61	140.14	181.00
0.3 sec S_a	120.94	140.44	181.97
1.0 sec S_a	57.70	71.83	100.14
Seattle			
PGA	33.77	48.61	76.49
0.2 sec S_a	75.20	113.63	161.34
0.3 sec S_a	62.25	103.36	145.47
1.0 sec S_a	22.06	32.23	55.97
St. Paul			
PGA	0.76	1.31	2.50
0.2 sec S_a	1.82	3.17	5.63
0.3 sec S_a	1.61	2.72	4.98
1.0 sec S_a	0.73	1.38	2.66
New York			
PGA	6.32	11.92	24.45
0.2 sec S_a	12.59	22.98	42.55
0.3 sec S_a	9.42	16.64	31.17
1.0 sec S_a	2.85	5.11	9.40
Memphis			
PGA	13.92	30.17	69.03
0.2 sec S_a	27.46	58.71	130.03
0.3 sec S_a	20.38	43.36	110.62
1.0 sec S_a	6.46	15.47	40.74

The equation describing the acceleration response spectrum, S_a, shown in Figure 2.6 can expressed as

$$S_a = 0.6\frac{S_{DS}}{T_0}T + 0.40S_{DS} \qquad \text{for } T < T_0,$$

$$S_a = S_{DS} \qquad \text{for } T_0 < T < T_s,$$

and

$$S_a = \frac{S_{D1}}{T} \qquad \text{for } T > T_s$$

(2.19)

The design spectra proposed by NCHRP Project 12-49 as adopted from NEHRP are based on the average response spectra developed by Frankel et al. (1997) from a large earthquake mapping project. Frankel et al. (1997) found that the level of confidence in the NEHRP spectra is related to the number of earthquakes recently observed as well as the knowledge of the type and locations of the faults in a particular region. For sites at which a large number of earthquakes were observed, the

COV is low; the COV is high for sites with few observed tremors. Frankel et al. (1997) provided maps showing uncertainty estimates for selected cities derived from Monte Carlo simulations. The data provided in this map show that the ratio between the 85th fractile and the 15th fractile for New York

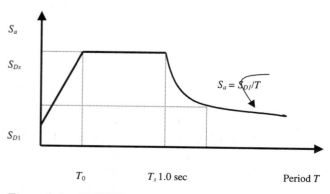

Figure 2.6.   NEHRP response spectrum.

City is on the order of 5. Assuming a normal distribution, this means that the COV would be on the order of 30%. For the Memphis area, the ratio of the two fractiles is about 3, resulting in a COV of about 25%. For San Francisco, the projected COV is about 15%, and for St. Paul and Seattle the projections are that the COVs would be about 40% and 25%, respectively. Frankel et al. (1997) also show that the mean value of the spectral accelerations is very close to the uniform hazard spectra they developed and that resulted in the NEHRP specifications. It is noted that these observations are within the range of the values reported by Seed, Ugas, and Lysmer (1976), who observed that the results of dynamic analyses using a variety of earthquake records resulted in a range of spectral responses with a COV of about 30% from the average spectra.

In summary, this study will assume that the mean spectral accelerations are equal to the design spectral accelerations proposed in NCHRP Project 12-49 (i.e., bias = 1.0) with a COV that depends on the frequency of earthquakes at the site. Thus, the COV varies between 15% for San Francisco and 40% for St. Paul. The spectral accelerations are also assumed to follow normal distributions.

Response Modification Factor

The response modification factor is related to the ductility of the system. The purpose of the response modification factor is to allow for a linear elastic analysis of structural systems although the system may exhibit large levels of plastic deformations. In fact, because earthquakes produce displacements rather than actual forces, traditional design of bridges for earthquakes allows bridge columns to reach their ultimate member capacity as long as the ductility capacity is not exceeded. To simplify the design and safety evaluation process, the ductility capacity, μ_{cap}, of a concrete column is modeled through a response modification factor, R. The response modification factor is used to allow the designer to perform a linear elastic analysis and to check the safety of the column members for a modified moment capacity equal to R_m times the actual moment capacity of the column. Hence, if the applied moment assuming linear elastic behavior is lower than R_m times the actual moment capacity, the column is considered to be safe.

Analytical research studies have concluded that the response modification factor R_m is directly related to the ductility capacity μ_{cap}. Thus, the response modification for a member, assuming an SDOF system, may be estimated if the member's ductility capacity, μ_{cap}, is known. Miranda (1997) found that for typical periods of bridge systems (0.5 to 1.5 sec) subjected to a representative sample of earthquake records, the response modification, R_m, is on the average equal to the bridge column ductility capacity ($R_m = \mu_{cap}$) with a COV of about 25%. This observation confirms the model first proposed by Newmark and Hall (1973) that was based on limited data from the El Centro Earthquake. The results of Miranda (1997) were calculated

for a variety of sites with a range of soil classifications. Liu et al. (1998) in a report to the National Center Earthquake Engineering Research (NCEER) and FHWA found that the COV reduces to about 17% if the earthquake records were chosen to match those that produce the design spectral accelerations.

In addition to the issue of the relation between R_m and μ_{cap}, another issue concerns the level of uncertainty associated with estimating the ductility capacity μ_{cap}. Results given by Priestley and Park (1987) show that the real ductility of bridge columns is on the average about 1.5 times higher than the ductility estimated from the design formulas with a COV of about 30%. Thus, using the results of Priestley and Park (1987) and Liu et al. (1998), the actual response modification factor, R_m, will be on the average 1.5 times the specified ductility capacity, $\mu_{specified}$, (bias of $R_m = 1.5$) with a COV of 34% $\left(0.90 \approx 1/\sqrt{1.25}\right)$. The probability distribution for R_m is assumed to be normal.

The last issue with the response modification factor concerns the range of values specified for use during the design process by AASHTO and other earthquake design codes. For example, it is noted that the response modification factor, R, specified by AASHTO for use during the analysis of single-column bents is set at 2.0, while $R_m = 3.5$ is used for multiple-column bents of essential structures. Also, values of 3.0 and 5.0 are used for "other structures." On the other hand, NCHRP Project 12-49 is proposing a single value, $R_m = 6.0$, for life safety for single-column and multicolumn bents and $R_m = 1.5$ to secure proper operation. ATC-6 mentions that an $R_m = 2.0$ is recommended for a wall-type pier "based on the assumption that a wall pier has low ductility capacity and no redundancy."

It is clear that the difference between the 2.0, 3.0, 3.5, 5.0, and 6.0 values of R_m used for the design of columns is not intended to account for the differences in the ductility capacities of the columns. Rather, the use of different values of R_m is meant to provide certain types of structures (particularly nonredundant and essential bridges) with higher levels of safety. In fact, since in all cases the design and construction procedures of columns in single-column bents or multicolumn bents are fairly similar, one would expect to find the ductility capacities of all columns to be about the same. It is noted that previous recommendations for the design of bridges under earthquake loads recommended that a response modification factor $R_m = 8$ be used. In addition, tests on bridge columns performed at the University of Canterbury (Zahn, Park, and Priestley, 1986) have shown that the ductility of properly confined columns can easily exceed 7.5, although some damage would be expected to occur.

Based on the information collected from the references mentioned above, this study will assume that the ductility level of confined concrete columns can be modeled by a response modification factor R that has an average value $\bar{R}_m$ of 7.5 and a COV of 34%. The probability distribution for R_m is assumed to be normal. It is noted that the values used herein are somewhat similar to those used by Hwang, Ushiba, and Shinozuka

(1988), who have recommended the use of a median value for R_m of 7.0 for shear walls with a COV of 40%. The use of a response modification factor during the analysis and design of bridge substructures implies that the design and safety evaluation account for the nonlinear capacity of the system. Thus, the reliability indexes calculated are for the system capacity rather than for member capacity.

Modeling Factor

The dynamic structural analysis produces a level of uncertainty in the final estimate of the equivalent applied forces and moments on a bridge substructure. These factors are due to the effects of lateral restraints from the slab, the uncertainty in predicting the tributary area for the calculation of mass, the point of application of the equivalent static load, the variability in soil properties and the uncertainty in soil classification, the effect of using a lumped mass model, the level of confidence associated with predicting the earthquake intensity, and so forth. Ellingwood et al. (1980) assumed that the modeling factor has a mean value equal to 1.0 and a COV on the order of 20% for buildings. The same value is used in these calculations.

Using the information provided in this section, the equivalent force applied on a bridge structure may be calculated using the following expression:

$$F_{apl} = \lambda_{eq} C' S_a(t'T) * \frac{A * W}{R_m} \qquad (2.20)$$

where

F_{apl} = the equivalent applied force;
λ_{eq} = the modeling factor for the analysis of earthquake loads on bridges;

C' = the response spectrum modeling parameter;
A = the maximum 75-year peak ground acceleration at the site (a 75-year design life is used in order to be consistent with the AASHTO LRFD specifications);
S_a = the calculated spectral acceleration, which is a function of the period T, where T is the bridge column period and the period modeling factor, t';
W = the weight of the system;
R_m = the response modification factor; and
C' and t' = modeling parameters that express the variation of the true spectrum from the design spectrum and the true natural period of the system from the design period.

The statistical data used in the reliability analysis for the random variables of Equation 2.20 are summarized in Table 2.5. The parameters not listed in Table 2.5 are assumed to be deterministic. It is noted that all the random variables except for the earthquake intensity are time independent in the sense that they are not affected by the exposure period. The earthquake intensity is time dependent because the longer a bridge's design life is, the higher the expected earthquake intensity that it will be exposed to. Finally, it is also noted that earthquakes last for short periods of time—a few seconds. In this study, for the sake of obtaining conservative results, it will be assumed that earthquakes will last about 30 sec ($\frac{1}{2}$ min), during which time the response of the bridge will continuously be at its maximum value. Table 2.5 shows the mean and COV for the maximum yearly earthquake acceleration intensities of the five sites depicted in Figure 2.4. A comparison between the COV of the earthquake intensities and the COVs of the other variables demonstrates the dominance of the uncertainties in estimating the maximum acceleration during the reliability analysis.

TABLE 2.5 Summary of input values for seismic reliability analysis

Variable		Bias	COV	Distribution Type	Reference
Earthquake modeling, λ_{eq}		1.0	20%	Normal	Ellingwood et al. (1980)
Spectral modeling, C'		1.0	Varies per site (15% to 40%)	Normal	Frankel et al. (1997)
Acceleration, A	San Francisco	1.83% g (yearly mean)	333%	from Figure 2.4	USGS website
	Seattle	0.89% g (yearly mean)	415%		
	Memphis	0.17% g (yearly mean)	1707%		
	New York	0.066% g (yearly mean)	2121%		
	St. Paul	0.005% g (yearly mean)	3960%		
Period, t'		0.90	20%	Normal	Chopra and Goel (2000)
Weight, W		1.05	5%	Normal	Ellingwood et al. (1980)
Response modification, R_m		7.5 (mean value)	34%	Normal	Priestley and Park (1987) and Liu et al. (1998)

2.4.3 Wind Load

As explained by Ellingwood et al. (1980), the wind load on bridge structures is a function of the expected wind speeds at the bridge site, pressure coefficients, parameters related to exposure and wind speed profile, and a gust factor that incorporates the effects of short gusts and the dynamic response of the structure. In general, the wind load (W) may be represented by an equation of the form

$$W = cC_p E_z G(\lambda_V V)^2 \tag{2.21}$$

where

c = an analysis constant,
C_p = the pressure coefficient,
E_z = the exposure coefficient,
G = the gust factor,
V = the wind speed at the desired return period, and
λ_V = the statistical modeling factor that accounts for the uncertainty in estimating V.

The exposure coefficient depends on the type of terrain (e.g., urban area, open country), and G depends on the turbulence of the wind and the dynamic interaction between the bridge structure and the wind. In addition to the time-dependent randomness of the wind speed, uncertainties in the estimation of the other terms contribute to the overall variability in the wind load. Ellingwood et al. (1980) assumed that c, C_p, E_z, and G are all random variables that follow normal distributions. Their mean values are equal to the nominal design values as calculated from design specifications, that is, the best estimates of each variable should be used during the design process. The COVs for each of the random variables are as follows: for c, the COV is 5%; for C_p, the COV is 12%; for G, the COV is 11%; and for E_z, the COV is 16%. The random variables represented by c, C_p, G, and E_z are time-independent, meaning that they do not change with the design life (or the exposure period of the structure to wind loads). Table 2.6 summarizes the data for the random variables considered.

For the purposes of this study, the fluctuations of the wind load applied on the structure with time are modeled as shown in Figure 2.7. The average time duration of each wind was found by Belk and Bennett (1991) to vary from 2.2 h to 5.62 h, depending on the site, with an average duration of 3.76 h and a COV close to 11%. This study assumes that each wind will

last about 4 h, during which time the wind force remains constant at its maximum value. Belk and Bennett (1991) have also found that an average site will be subjected to about 200 winds per year, which is the value that is used in this project to study the load combination problem. For the sake of simplicity, the model used herein assumes independence between individual wind events. The statistical models for wind speed are discussed further below.

Wind Speeds

Data on wind velocities, extreme wind events, and the effect of winds on structures are available in the engineering literature (e.g., see Whalen and Simiu, 1998). However, much of this work is still under development and has not been applied in structural design codes. For this reason, in this project we will focus on the work of Ellingwood et al. (1980), which led to the development of ANSI 58 and subsequently the ASCE 7-95 guidelines. The AASHTO LRFD specifications are also based on the models developed for ASCE 7-95. According to the AASHTO LRFD, the basic design wind speed of 160 km/h (100 mph) is associated with an annual probability of 0.02 of being exceeded (i.e., the nominal recurrence interval is about 50 years). This value is for an open terrain with scattered obstructions having heights generally less than 9.1 m (30 ft) (ASCE 7-95 Exposure C). According to the ASCE 7-95 basic wind speed map, the 160 km/h wind would be the expected maximum wind for a 50-year return period for East Coast regions such as Philadelphia and Washington, D.C. While interior regions would have a design wind speed of 145 km/h (90 mph), coastal regions such as Long Island and Rhode Island would require a design speed of 190 km/h (120 mph), whereas regions such as New York City and Boston would require 180 km/h (110 mph) design winds. The higher design wind speeds for the coastal regions are due to the occurrence of hurricanes. The design wind speed maps do not correspond to actual projected 50-year maximum wind speed but give conservative envelopes to the expected 50-year wind speeds. In fact, based on data collected by Simiu, Changery, and Filliben (1979) for annual maximum wind speeds and assuming that wind speeds follow a Gumbel Type I extreme value distributions, Ellingwood et al. (1980) developed the results shown in Table 2.7 for wind speeds at different sites in the interior region of the United States. Belk and Bennett (1991) performed a statistical analysis of the data col-

TABLE 2.6 Summary of input values for wind reliability analysis

Variable	Bias	COV	Distribution Type	Reference
Analysis factor, c	1.0	5%	Normal	Ellingwood et al. (1980)
Pressure coefficient, C_p	1.0	12%	Normal	Ellingwood et al. (1980)
Exposure coefficient, E_z	1.0	16%	Normal	Ellingwood et al. (1980)
Gust factor, G	1.0	11%	Normal	Ellingwood et al. (1980)
Wind velocity, V	from Table 2.7	from Table 2.7	Gumbel	Simiu et al. (1979)
Statistical variable, λ_V	1.0	7.5%	Normal	Simiu et al. (1979)

36

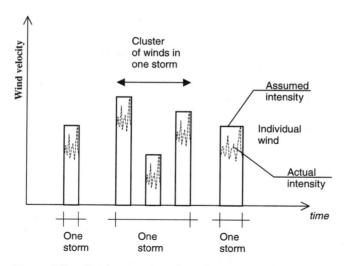

Figure 2.7.   Representation of wind events.

lected by Simiu, Changery, and Filliben (1979) and confirmed that wind speeds in the interior are best modeled by a Gumbel distribution. Figure 2.8 gives a curve adapted from Belk and Bennett (1991) that shows the close fit between the Gumbel Type I distribution and the data collected by Simiu, Changery, and Filliben (1979) for yearly maximum wind speeds. In subsequent work, Simiu, Heckert, and Whalen argued that regular wind speeds in interior regions are best modeled as reverse Weibull distributions (Simiu and Heckert, 1995; Whalen and Simiu, 1998). However, describing the reverse Weibull distribution requires three statistical parameters. These three parameters are the mean, the standard deviation, and a cut-off threshold value that describes the maximum tail length of the distribution. Simiu and Heckert (1995) argue that "it is difficult to provide reliable quantitative estimates of the tail length parameters" because of the fluctuations in the available estimates of the cut-off threshold values. Hence, because it is difficult to use the reverse Weibull and until the data on the threshold value are made available, in this study a Gumbel distribution is used to model the wind speeds. The Gumbel assumption is

compatible with the work conducted on the calibration of the ASCE design load criteria. Table 2.7 provides the statistical data for the wind speed of typical U.S. sites as assembled by Ellingwood et al. (1980) based on the work of Simiu, Changery, and Filliben (1979). This data will be used in this study to perform the reliability analysis of bridges subjected to wind loads and the combination of winds and other extreme events.

In addition to the wind speed, Ellingwood et al. (1980), following Simiu, Changery, and Filliben (1979), suggest that the statistical uncertainties caused by the limitations in the number of data points used to find the statistics of the wind speed should be included during the reliability analysis. Ellingwood et al. (1980) found that the statistical uncertainties would produce a COV on the order of 7.5% in the projection of the results to the large period of time. The statistical uncertainties are modeled through the variable λ_V of Equation 2.21. λ_V has a mean equal to 1.0 and a COV equal to 7.5% and follows a normal distribution.

The emphasis of this study is on regular windstorms in interior regions of United States excluding the effect of hurricanes and tornados because of the lack of sufficient data on the effects of these strong winds on typical bridge configurations.

2.4.4 Scour

The HEC-18 design equation for local scour around bridge piers is a function of the flow depth, the pier nose shape, the angle of attack of the flow, the streambed conditions, the soil material size, the diameter of the pier, and the Froude number. The equation that is intended to predict the depth of maximum scour for design purposes is as follows:

$$y_{max} = 2y_0 K_1 K_2 K_3 K_4 \left(\frac{D}{y_0}\right)^{0.65} F_0^{0.43} \qquad (2.22)$$

where

y_{max} = the maximum depth of scour;
y_0 = the depth of flow just upstream of the bridge pier excluding local scour;

TABLE 2.7   Wind load data in mph based on Ellingwood et al. (1980)

Site	Annual Maximum		50-Year Maximum	
	Mean Annual Velocity (mph)	COV Annual Velocity	Mean 50-Year Wind Velocity (mph)	COV 50-Year Wind Velocity
Baltimore, MD	55.9	0.12	76.9	0.09
Detroit, MI	48.9	0.14	69.8	0.10
St. Louis, MO	47.4	0.16	70.0	0.11
Austin, TX	45.1	0.12	61.9	0.09
Tucson, AZ	51.4	0.17	77.6	0.11
Rochester, NY	53.5	0.10	69.3	0.08
Sacramento, CA	46.0	0.22	77.4	0.13

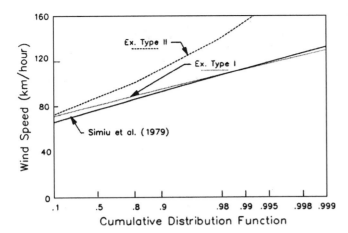

Figure 2.8. Comparison of cumulative probability distribution fit (after Belk and Bennett, 1991).

K_1, K_2, K_3, and K_4 = coefficients that account for the nose shape of the pier, the angle between the direction of the flow and the direction of the pier, the streambed conditions, and the bed material size;

D = the pier diameter; and

F_0 = Froude number, which is defined as

$$F_0 = \frac{V}{(gy_0)^{0.5}} \qquad (2.23)$$

where V is the mean flow velocity at the pier, and g is the acceleration due to gravity.

The flow depth, y_0, and flow velocity, V, are related to the flow discharge rate, Q, at a given point in time and the shape of the channel represented by the cross sectional area of the stream, A_0. This relationship is given by

$$Q = A_0 V. \qquad (2.24)$$

For a rectangular cross section with a constant width, b, and flow depth, y_0, the relationship becomes

$$Q = by_0 V. \qquad (2.25)$$

The relationship between the flow velocity, V, and the hydraulic radius that is related to the flow depth, y_0, can be expressed using Manning's equation:

$$V = \frac{\Phi_0}{n} R_H^{2/3} S_0^{1/2} \qquad (2.26)$$

where

n = Manning roughness coefficient,
R_H = the hydraulic radius,

S_0 = the slope of the bed stream, and
Φ_0 = a unit adjustment parameter equal to 1.486 when using U.S. units (ft and sec).

For SI units, Φ_0 equals 1.0. For a rectangular channel of width b, the hydraulic radius, R_H, can be calculated by

$$R_H = \frac{by_0}{b + 2y_0}. \qquad (2.27)$$

Thus, the relationship between the flow rate Q and the flow depth y is given from the equation

$$Q = by_0 \frac{\Phi_0}{n} \left(\frac{by_0}{b + 2y_0} \right)^{2/3} S_0^{1/2}. \qquad (2.28)$$

Typical values for Manning's roughness coefficient, n, vary from 0.025 to 0.035 for earth (respectively for good condition and for weeds and stones). It is noted that estimating the appropriate Manning roughness coefficient is associated with a high level of uncertainty. Researchers at Hydraulic Engineering Center (1986) determined that the roughness coefficient, n, follows a lognormal distribution with a COV ranging from 20% to 35% (with an average value of 28%). Therefore, in this report, n is taken as a random variable with a bias equal to 1.0 compared with the recommended tabulated values and a COV equal to 28%. It is also assumed that the slope S is known and, thus, the uncertainties in V are primarily due to the uncertainties in estimating n.

Different researchers and organizations have used different probability distribution types to model the discharge rate, Q. Extreme Type I distributions, lognormal distributions, and logPearson distributions are most commonly used. By studying several probability plots, the maximum yearly discharge rates for different rivers were found to follow a lognormal distribution. An example of the fit on lognormal probability paper is presented in Figure 2.9 for the Schohaire Creek. Data from five different rivers are also summarized in Table 2.8. The raw data that generated the statistics shown in Table 2.8 were collected from the USGS website. Knowing the probability distribution for the yearly discharge rate, the maximum 75-year flood discharge has a cumulative probability distribution, $F_{Q75}(x)$, related to the probability distribution of the 1-year maximum discharge by

$$F_{Q75}(x) = F_Q(x)^{75}. \qquad (2.29)$$

Equation 2.29 assumes independence between the floods observed in different years. This assumption is consistent with current methods for predicting maximum floods.

The estimation of the mean and standard deviation of the 75-year maximum discharge rate (Q_{75}) for each river, as shown in Table 2.8, is associated with some level of uncertainty. This level of uncertainty depends on the number of samples available to calculate the means and standard deviations. The

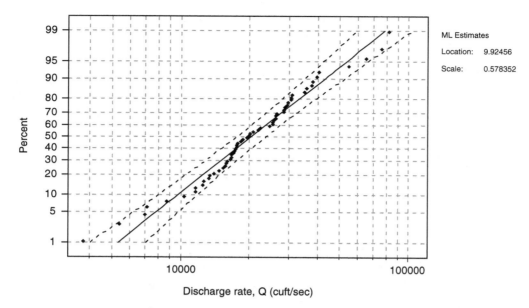

Figure 2.9. Representation of discharge rate for Schohaire Creek on lognormal probability plot.

analysis of the data provided in the USGS website for the five rivers analyzed in this report show that the mean values provided in Table 2.8 are within a 95% confidence level. For this reason, a modeling variable λ_Q is used in this report to express the variability in the prediction of Q. This modeling variable is thus assumed to follow a normal (Gaussian) distribution and to have a bias of 1.0 and a COV of 5%.

The HEC-18 approach has been extensively used for practical design considerations although the HEC-18 empirical model provides conservative estimates of scour depths and is known to have the following five limitations:

1. The HEC-18 equation is based on model scale experiments in sand. In a recent evaluation against full-scale observations from 56 bridge sites, HEC-18 has been found to vastly over-predict the scour depth (Landers and Mueller, 1996). A comparison of the HEC-18 equation and the measured depths are illustrated in Figure 2.10, which is adapted from Landers and Mueller (1996).

2. Once a flood begins, it takes a certain period of time for the full extent of erosion to take effect. Thus, if the flood is of a short duration, the maximum scour depth may not be reached before the flood recedes. On the other hand, prior floods may have caused partial erosions accelerating the attainment of the maximum scour depth. HEC-18 does not predict the length of time required for the maximum scour depth to be reached and assumes that the maximum depth is always reached independent of the flood duration and the level of scour incurred by prior floods.

3. The HEC-18 model does not distinguish between live-bed scour and clear-water scour in terms of the time required to reach equilibrium scour depth and the differences in the expected magnitudes of scour depths for these different phenomena (Richardson and Davis, 1995).

4. HEC-18 was developed based on experimental data obtained for sand materials. Some work is in progress to study the applicability of HEC-18 for both sand and

TABLE 2.8 Probability models for five rivers

River	Log Q	$S_{\log Q}$	D_n*	Average 75-year Q, Q_{75} (ft³/sec)	COV of Q_{75}
Schohaire	9.925	0.578	0.067	85,000	29%
Mohawk	9.832	0.243	0.068	34,000	12%
Sandusky	9.631	0.372	0.086	38,000	18%
Cuyahoga	9.108	0.328	0.065	20,000	16%
Rocky River	9.012	0.378	0.049	21,000	19%

*D_n is the K–S maximum difference between the measured cumulative probability and expected probability value. More than 60 data points were available for each of the five rivers. The D_n values obtained indicate that the lognormal distribution is acceptable for a significant level $\alpha = 20\%$.

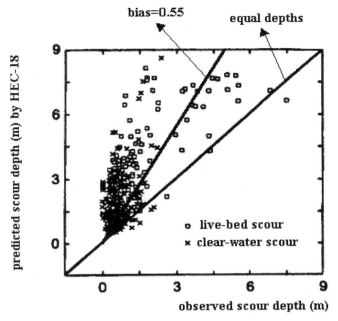

Figure 2.10. *Comparison of HEC-18 predictions to observed scour depths (based on Landers and Mueller, 1996, Figure 2).*

clay streambed materials because it is well known that these materials behave differently (Briaud et al. 1999).

5. The usual assumption is that scour is deepest near the peak of a flood but may be hardly visible after floodwaters recede and scour holes refill with sediment. However, there are no known methods to model how long it takes a river to backfill the scour hole. Refill can occur only under live-bed conditions and depends on the type and size of the transported bed material (sand or clay). Even when refill occurs, it will take a considerable length of time for the refill material to sufficiently consolidate and restore the pier foundation to its initial strength capacity. Although such information is not precisely available, a number of bridge engineers have suggested that periods of 2 to 3 months are reasonable for sandy materials with longer periods required for clays.

Based on the observations made above, a possible reliability model for the reliability analysis of a bridge pier under the effect of scour is proposed based on the work of Johnson as presented in Appendix B. This model is based on the observation that Equation 2.22 gives a safe value for the depth of scour. The average ratio of the observed scour depth compared with the HEC-18 predicted depth was found by Johnson (1995) to be about 0.55 (bias value is 0.55). Also, the ratio of the true scour value over the predicted value has a COV of 52%. This ratio is represented by a scour modeling variable, λ_{sc}, that is assumed to follow a normal distribution. Thus, according to Johnson (see Appendix B), the true scour value can be represented by an equation of the form

$$y_{\max} = 2\lambda_{sc}y_0 K_1 K_2 K_3 K_4 \left(\frac{D}{y_0}\right)^{0.65} F_0^{0.43}. \quad (2.30)$$

It should be noted that Johnson (1995) also recommends that the factor K_3 representing the effect of streambed condition be treated as a random variable with a bias equal to 1.0 and a COV equal to 5% to account for the possible variability of the streambed between floods. Table 2.9 summarizes the input data for the random variables appropriate for use with Equation 2.30. As indicated above, Q is a time-dependent variable while the remaining random variables are time-independent. Equation 2.30 assumes that the statistical properties of the modeling variable, λ_{sc}, are constant for all floods and rivers. However, a review of the Landers and Mueller (1996) data plotted in Figure 2.10 has revealed that this is not necessarily the case. For this reason, an alternate model has been developed in Appendix I and is described in the next paragraph.

Equation 2.30 provides a reliability model that is compatible with the models proposed by researchers for the live load, earthquake, and wind events. However, a careful review of the data depicted in Figure 2.10 shows that for low scour depths, λ_{sc} appears higher than the average value of 0.55 and for high scour depths, λ_{sc} appears to be lower. For this reason, an alternate reliability model is proposed based on the regression analysis of the data published by Landers and

TABLE 2.9   Input data for reliability analysis for scour alone

Variable	Mean Value	COV	Distribution Type	Reference
Discharge rate, Q	from Table 2.8	from Table 2.8	Lognormal	Based on USGS website
Modeling variable for Q, λ_Q	1.0	5%	Normal	Based on USGS website
Modeling variable, λ_{sc}	0.55	52%	Normal	Johnson (1995)
Manning roughness, n	0.025	28%	Lognormal	Hydraulic Engineering Center (1986)
Bed condition factor, K_3	1.1	5%	Normal	Johnson (1995)
Residual error, ε	0.0	Standard deviation = 0.406	Normal	Appendix I

40

Mueller (1996). Based on the regression fit as described in Appendix I, the observed scour depths around rounded bridge piers founded on non-cohesive soils may be modeled by an equation of the form

$$\ln y_{max} = -2.0757 + 0.6285 \ln D + 0.4822 \ln y_0 + 0.6055 \ln V + \varepsilon \qquad (2.31)$$

where

D = the pier diameter,
y_0 = the flow depth,
V = the flow velocity, and
ε = the residual error.

As explained above, D is a deterministic variable because the pier diameter can be accurately known even before the actual construction of the pier. y_0 and V are random variables that depend on Manning's roughness coefficient, n, and the 75-year maximum discharge rate, which are random variables having the properties listed in Table 2.9. Based on the analysis of the residuals affected in Appendix I, ε may be considered to follow a normal distribution with mean equal to zero and a standard deviation equal to 0.406.

Finally and as mentioned above, it is noted that under live bed conditions, the local scour hole is normally assumed to refill as the scour-causing flood recedes. However, the available literature does not provide precise information on how long it normally takes for the foundation to regain its original strength. This is believed to depend on the type of material being deposited by live-bed streams. For example, fine sands may tend to regain their strengths within a short period of time (perhaps 2 to 3 months). On the other hand, cohesive materials such as clays may take much longer to consolidate and regain their original strengths. As a compromise, the calculations performed in this study will assume that it takes about 6 months (½ year) for a foundation to regain its original strength. It is further assumed that the scour depth produced by the maximum yearly flood will remain at its maximum value for this half-year period. This assumption will also indirectly account for the effects of smaller floods within that period of time. The proposed model will also assume that the scour depth will be reached instantaneously as the flood occurs and that the flooding period is always long enough for the maximum scour depth to be reached.

2.4.5 Vessel Collision Forces

Considerable effort was spent on studying vessel collision forces during the development of AASHTO's *Guide Specifications and Commentary for Vessel Collision Design of Highway Bridges* (1991). The AASHTO guide uses a reliability formulation following the recommendations made by several International Association for Bridge and Structural Engineers (IABSE) workshops and symposia. An IABSE

Working Group assembled a state-of-the-art report summarizing the findings of an international group of researchers (Larsen, 1993). The latter document gives an overview of the background information that led to the development of the AASHTO guide specifications. The guide gives an example outlining the application of the guide's Method II, which gives a probability-based analysis procedure for determining the design forces caused by ship impacts with bridges. In addition, Whitney et al. (1996) describe the application of the AASHTO vessel collision model for barge traffic over the Ohio River. The model used in this study to perform the reliability analysis of bridge piers subjected to vessel collisions is based on the guide's Method II and follows the example described by Whitney et al. (1996). The calculations performed for this report are for barge collision forces such as those that may occur at the piers of bridges spanning rivers with heavy barge traffic. Bridges spanning waterways with big ship traffic are normally unique structures that should be studied on an individual basis.

Based on the AASHTO guide method, the design barge collision force can be represented by an empirically derived equation as a function of the barge bow damage depth. The design force equation takes the form

$$P_B = 60a_B \qquad \text{for } a_B < 0.1\text{m},$$

and $\qquad (2.32)$

$$P_B = 6 + 1.6a_B \qquad \text{for } a_B \geq 0.1\text{m}$$

where P_B is the nominal design force in MN (meganewtons), and a_B is the barge bow damage depth.

According to the AASHTO guide specifications, the barge bow depth is calculated from the kinetic energy of the moving barge. Because barges in large rivers travel in flotillas, the kinetic energy that should be used in calculating the collision force should account for the masses of all the barges in a column of the flotilla when head-on collisions are considered. Hence, the kinetic energy KE is calculated as

$$KE = \frac{C_H W V^2}{2 \times 9.81} \qquad (2.33)$$

where

W = the weight of a flotilla column,
V = the speed of the flotilla at impact,
9.81 = the acceleration due to gravity in m/s^2, and
C_H = a hydrodynamic coefficient that accounts for the effect of the surrounding water upon the moving vessel.

As an example, Whitney et al. (1996) suggest that the value $C_H = 1.05$ is appropriate for the Ohio River because of the relatively large underkeel clearance and accelerations in the direction of the ship length (i.e., there are no large lateral motions as those associated with barge berthing). The barge damage depth, a_B, is given as

$$a_B = \frac{\left[\sqrt{1.00 + 0.13KE} - 1\right] \times 3.1}{R_B} \qquad (2.34)$$

where R_B is the correction factor for barge width given as $R_B = B_B/10.68$ where B_B is the barge width in meters (or $R_B = B_B/35$ in feet for U.S. units). The correction factor is meant to account for the difference between the width of the barge tested to empirically obtain the barge damage depth equation and the barge width of the impacting vessel.

The kinetic energy of Equation 2.33 must be calculated based on the speed at impact that must account for the speed of the flotilla relative to the river and the river flow speed. When the main piers of a bridge are adjacent to the vessel transit path, the transit speed is used for the relative speed of impact. Otherwise, AASHTO gives an empirical equation that describes how the speed varies from the travel speed to the river flow speed as a function of the distance between the transit path and the pier location. For example, Whitney et al. (1996) found that the flotilla speed in the Ohio River may reach up to 3.13 m/sec (10.3 ft/sec). Given that the river speed is on the average 1.86 m/sec (6.1 ft/sec), the speed at impact will be equal to 4.99 m/sec (16.4 ft/sec).

Modeling Factor

Equation 2.32 for the nominal impact force was developed based on experimental data of individual barge collisions with lock entrance structures and bridge piers. These studies included dynamic loading with a pendulum hammer, static loading on barge models, and numerical computations. However, the tests were conducted for single barges at low velocities and not multibarge flotillas traveling at high velocities. Whitney et al. (1996) report that the actual crushing depths as observed from accidents on the Ohio River were much lower than those calculated from the results of the AASHTO equations. This may be due to the significant energy loss that occurs among the barges of the flotilla because of friction and crushing. To correct for the differences between the calculated damage and the observed damage, the AASHTO guide uses a modeling variable, x. Thus, the actual impact force is given as

$$P = xP_B \qquad (2.35)$$

where x is the modeling variable, and P_B is the predicted value of the impact force that is given by Equation 2.32.

The random variable, x, gives the ratio of the actual impact force P to the nominal impact force P_B. A probability density

and a cumulative distribution function are given to describe x, as shown in Figure 2.11 (which is adapted from Figure C4.8.3.4-2 of the AASHTO guide [1991]). For a given value of barge weights in a flotilla column, W, P_B is calculated from Equations 2.32 through 2.34. The probability that P is greater than a certain fraction of P_B is obtained from Figure 2.11(b). Figure 2.11 shows that P_B gives a very conservative estimate of the impact force. For example, Figure 2.11(b) shows that the probability that P is greater than $0.1P_B$ (or $x = P/P_B = 0.1$) is only 0.1 (or 10%). The probability that P is greater than $0.5P_B$ (or $P/P_B = 0.5$) is 0.0556 (or 5.56%). The probability that P is greater than P_B ($P/P_B = 1$) is zero. The AASHTO guide states that the results illustrated in Figure 2.11 were obtained from the work of Cowiconsult (1987) for ship collisions. The same curve was assumed by Whitney et al. (1996) to be valid for collisions of barge flotillas.

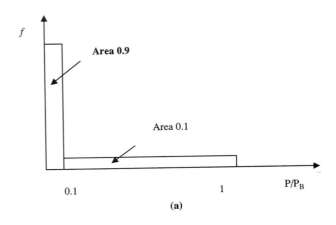

(a)

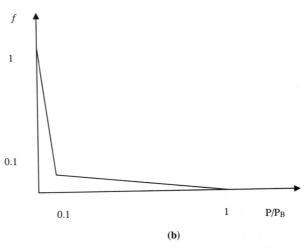

(b)

Figure 2.11. Distribution function for vessel collision force modeling variable: (a) probability density function for relative magnitude of the collision force, x; (b) distribution function for x = P/P_B exceeding a given level.

42

Rate of Collisions

Equation 2.35 above gives the force at impact given that a barge column with a known weight and speed has collided with a bridge pier. However, because not all flotillas are expected to collide, the reliability calculations must account for the number of collisions expected during the design life of the bridge structure. Although the AASHTO guide specifications develop the design criteria in terms of annual risk, to be consistent with the AASHTO LRFD requirements, bridges should be designed to satisfy a minimum level of reliability for a 75-year design life.

As presented in the AASHTO guide specifications and the IABSE report, the annual failure rate caused by vessel collisions, *AF*, can be expressed as

$$AF = \sum_i N_i \, PA_i \sum_k PG_{i,k} \, PC_{i,k} \qquad (2.36)$$

where

N_i = the number of vessels (or flotillas) of type *i* that cross the waterway under the bridge in 1 year;

PA_i = the probability of vessel aberrancy (of straying away from normal navigation channel) for vessels of type *i*;

$PG_{i,k}$ = the geometric probability of collision of ship type *i* with bridge element *k* (this gives the probability of having a collision with bridge member *k*, given that an aberrancy occurred in ship or a flotilla of type *i*); and

$PC_{i,k}$ = the probability that the bridge will collapse given that a vessel of type *i* has collided with member *k* of the bridge.

Equation 2.36 leads to the yearly rate of collisions for each vessel (or flotilla) of type *i* into a particular bridge element, *k*, as

$$\nu_i = N_i \, PA_i \, PG_i \, . \qquad (2.37)$$

Below is a description on the method proposed by the AASHTO guide to calculate the probability of aberrancy and the geometric probability. Using the rate of collisions and the frequency distribution of vessel (or flotilla types), the probability distribution of the force P_B of Equation 2.32 can be calculated as explained further below.

Probability of Aberrancy, PA_i

The probability of aberrancy (sometimes referred to as the causation probability) is a measure of the risk of a vessel losing control as a result of pilot error, adverse environmental conditions, or mechanical failure. The AASHTO guide states that the evaluation of accident statistics indicates that human error (causing 60% to 85% of the aberrancy cases) and envi-

ronmental conditions form the primary reasons for accidents. The environmental causes include poor visibility, strong currents, winds, channel alignment, and so forth. The IABSE report (Larsen, 1993) states that statistical data in major waterways show that the probability of vessel aberrancy varies from about 0.5 to 7 in 10,000 passages. Worldwide, the average is about 0.5 in 10,000 passages. Because such data are particularly hard to obtain for new bridge sites, the AASHTO guide proposes an empirical equation based on historical accident data. The AASHTO equation (Equation 4.8.3.2-1 in the AASHTO guide [1991]) accounts for the following factors: (1) the geometry of the navigation channel and the location of the bridge in the channel (turns and bends); (2) the current direction and speed; (3) the crosscurrents; and (4) vessel traffic density. The equation is given as

$$PA = BR_a(R_B)(RC)(R_{XC})(R_D) \qquad (2.38)$$

where

PA = probability of aberrancy;

BR_a = aberrancy base rate = 0.6×10^{-4} for ships or 1.2×10^{-4} for barges;

R_B = correction factor for bridge location = 1.0 for straight paths regions (varies as function of angle θ for vessel paths with turns and bends);

R_c = correction factor for current acting parallel to vessel path;

R_{xc} = correction factor for crosscurrents acting perpendicular to vessel transit path; and

R_D = correction factor for vessel traffic density depending on the frequency of vessels meeting, passing, or overtaking each other in the immediate vicinity of the bridge.

For example, the actual data collected by Whitney et al. (1996) for barge collisions in the Ohio River shows that the rate of aberrancy has an average value of 5.29×10^{-4}, which is higher than the 1.20×10^{-4} AASHTO value. They also found that the rate of aberrancy was equal to 13.78×10^{-4} for the Tennessee River, 18.11×10^{-4} for the Cumberland River, 3.14×10^{-4} for the Green River, and 1.2×10^{-4} for the Kentucky River. The IABSE report (Larsen 1993) indicates that the probability of collision to bridge piers increases by a factor of 3 when the wind speeds are in the range of 40 to 50 km/h (25 to 30 mph) as compared with the aberrancy rates when the wind speed are 20 to 30 km/h (12 to 19 mph).

Geometric Probability, PG_i

The geometric probability is defined as the probability of a vessel hitting the bridge pier given that the vessel has lost control. This probability is a function of many parameters, including the geometry of the waterway, the location of bridge piers, the characteristics of the vessel, and so forth. The AASHTO guide specification developed an empirical approach for finding the geometric probability. The AASHTO approach is based on the following three assumptions:

1. The lateral position of a vessel in the waterway follows a normal distribution with a mean value centered on the required path line (centerline of navigation route).
2. The standard deviation of this lateral position distribution is equal to the overall length of vessel designated as *LOA*. In the case of flotillas, Whitney et al. (1996) recommend using the total length of the flotilla (i.e., barge length times number of barges in a column).
3. The geometric probability is calculated from the normal distribution depending on the location of the pier relative to the centerline of the navigation route, the width and orientation of the pier, and the width of the vessel. For flotillas, the total width of the flotilla should be used.

The method to calculate the geometric probability, *PG*, is illustrated in Figure 2.12, which is adopted from the AASHTO guide and the IABSE report. The use of a standard deviation equal to *LOA* was justified based on accident data to reflect the influence of the size of the colliding ship.

Probability Distribution of the Predicted Impact Force, P_B

The force P_B of Equation 2.32 depends on the type of impacting vessel (or flotilla), including the weight of the vessel, its length, and other geometric features. Given the statistical data on the types of vessels (or flotillas) and their properties, the probability distribution of the predicted impacting force P_B can be assembled. The data on the type of vessels and their weights can be gathered from agencies that track the traffic in U.S. waterways such as the U.S. Army Corps of Engineers (USACE), which has provided data on barge traffic in

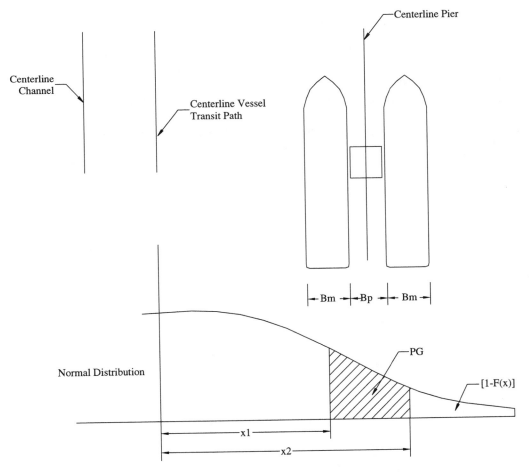

Geometric Probability of Vessel Collision with the Main Pier

Figure 2.12. Model for geometric probability of vessel collision with main bridge pier [based on the AASHTO guide (1991) and the IABSE report (Larsen, 1993)].

44

the Mississippi River near Memphis, Tennessee. Figure 2.13 shows the yearly probability distribution function for the impacting force calculated for the Mississippi River based on the USACE's data. More details explaining how the distribution function was assembled are provided in Appendix D.

The AASHTO guide specifies that when a vessel collides with a bridge pier, the impact force, P_B, obtained from Equation 2.32, will be applied as a concentrated force at the high water level. Several possible modes of failure may result because of the application of the force, P_B. The most likely failure modes include shear failure of the pier at the point of impact, moment failure of the pier or the foundation, and soil failure in the pier foundation.

In summary, the model used in this report to study the reliability of a bridge pier subjected to vessel collisions is based on the model proposed by Whitney et al. (1996), which follows the AASHTO model (1991). The process followed consists of the following steps and assumptions:

- The geometric probability, PG_i, of a flotilla type i colliding with the bridge pier depends on the flotilla overall length, LOA_i, as described in Figure 2.12. Each flotilla of type i may have a different length depending on the number of barges in each column and the length of each barge. To simplify the problem, it is herein assumed that the number of barges in a column of a flotilla is equal to the average number of barges as reported by Whitney et al. (1996).

- The expected number of collisions of flotillas of type i with the pier is equal to $N_i PA_i PG_i$ where N_i is the number of flotillas of type i crossing the site, PA_i is the probability of aberrancy of flotilla type i, and PG_i is the geometric probability of collision of flotilla type i.

- The nominal force applied on the pier if a flotilla of type i collides with the bridge pier is calculated from Equation 2.31 if the total weight of the flotilla and the width of the impacting barge are known. Each flotilla of type i may have a different total weight depending on the number of barges in the flotilla and the weight of each barge. Whitney et al. (1996) use the weight of one column of barges to find the kinetic energy at impact. The assumption is that the other columns are loosely tied to the impacting column such that at impact, only the barges in one column will contribute to the impact energy. To simplify the problem, it is herein assumed that the weight is equal to the average number of barges in one column times the average weight of the barges that form the flotilla. Average values for number of barges in a column and barge weights are provided by Whitney et al. (1996).

- By assembling the nominal "average" impact force for each flotilla type and the expected number of collisions for each flotilla type, a cumulative distribution of the yearly impact force can be drawn as is shown in Figure 2.13.

- Equation 2.14 can then be used to find the probability distribution for the 75-year bridge design life or other return periods.

Because P_B is calculated using the average vessel weight for each vessel type, it is herein proposed to account for the probability of having vessels heavier than the average value. This is achieved by including a weight-modeling factor during the reliability calculations. The data provided by

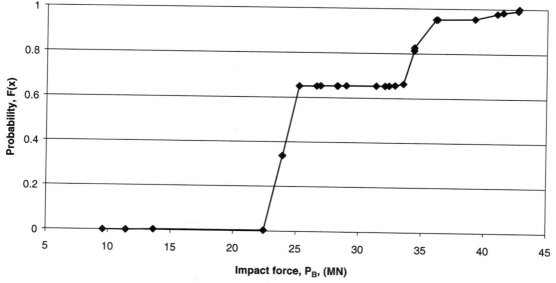

Probability distribution for yearly impact force

Figure 2.13. Cumulative distribution function for calculated ship collision force, P_B.

USACE show that the weight-modeling variable has a mean value of 1.0 and a COV of 10%. It is assumed to follow a normal distribution following the suggestion of Whitney et al. (1996). This weight-modeling factor is used to supplement the force-modeling factor x such that Equation 2.35 is modified to become

$$P = xwP_B. \qquad (2.35')$$

The three random variables—x, w, and P_B—used in modeling the final load effects of impact vessels are summarized in Table 2.10.

2.5 RISK ASSESSMENT MODELS FOR LOAD COMBINATIONS

As mentioned in Chapter 1, the Ferry Borges–Castanheta model is used in this report to study the reliability of a bridge system subjected to a combination of load events. In this section, the Ferry Borges–Castanheta model is described for two load processes and is illustrated in Figure 2.14 (Turkstra and Madsen, 1980; Thoft-Christensen and Baker, 1982). The model assumes that each load effect is formed by a sequence of independent load events, each with an equal duration. The service life of the structure is then divided into equal intervals

TABLE 2.10   Summary of random variables for flotilla collision with bridge pier

Variable	Symbol	Mean	COV	Distribution Type	Reference
Applied nominal force	P_B	from Figure 2.13	from Figure 2.13	from Figure 2.13	Based on AASHTO (1991) and Whitney et al. (1996)
Force-modeling factor	x	from Figure 2.11	from Figure 2.11	from Figure 2.11	AASHTO (1991)
Weight-modeling variable	w	1.0	10%	Normal	Based on data from USACE

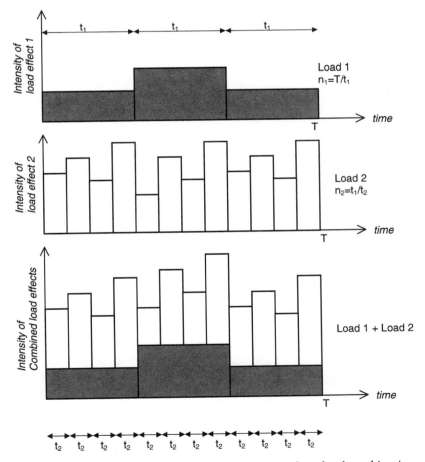

Figure 2.14.   Description of Ferry Borges–Castanheta load combination model.

46

of time, each interval being equal to the time duration of Load 1, t_1. The probability of Load 1 occurring in an arbitrary time interval can be calculated from the occurrence rate of the load. Simultaneously, the probability distribution of the intensity of Load 1 given that the load has occurred can be calculated from statistical information on load intensities. The probability of Load 2 occurring in the same time interval as Load 1 is calculated from the rate of occurrence of Load 2 and the time duration of Loads 1 and 2. After calculating the probability density for Load 2 given that it has occurred, the probability of the intensity of the combined loads can be easily calculated. The Ferry Borges–Castanheta model can be summarized with reference to Figure 2.14 as follows:

1. The intensity of the effect of Load 1 is a function of time t, defined as x_1.
2. The intensity of the effect of Load 2 is a function of time t, defined as x_2.
3. x_1 and x_2 vary randomly in function of time as shown in Figure 2.14.
4. Every time Load 1 occurs, it lasts for a fixed time duration equal to t_1.
5. Every time Load 2 occurs, it lasts for a fixed time duration equal to t_2.
6. The life span of a bridge structure, T, is an integer multiple of t_1.
7. t_1 is an integer multiple of t_2.
8. Load 1 occurs a total of n_1 times in the life of the bridge ($T = n_1 \times t_1$).
9. Load 2 occurs a total of n_2 times when Load 1 is on ($t_1 = n_2 \times t_2$).
10. Each occurrence of Load 1 is independent from the previous occurrences.
11. Each occurrence of Load 2 is independent from the previous occurrences.
12. Loads 1 and 2 are also independent of each other.
13. The combined load effect is defined as $X = x_1 + x_2$.

The load combination problem consists of predicting the maximum value of X, $X_{\max,T}$, that is likely to occur in the lifetime of the bridge, T. In this lifetime there will be n_1 independent occurrences of the combined load effect, X. The maximum value of the n_1 possible outcomes is represent by

$$X_{\max,T} = \max_{n_1}[X] \tag{2.39}$$

The maximum value of x_2 that is likely to occur within a time period t_1 (when Load 1 is on) is defined as $x_{2\max,t1}$. Since Load 2 occurs n_2 times within the time period t_1, $x_{2\max,t1}$ is represented by

$$X_{2\max,t1} = \max_{n_2}[x_2] \tag{2.40}$$

$X_{\max,T}$ can then be expressed as

$$X_{\max,T} = \max_{m}[x_1 + x_{2\max,t1}] \tag{2.41}$$

or

$$X_{\max,T} = \max_{n_1}\left[x_1 + \max_{n_2}[x_2]\right] \tag{2.42}$$

The problem will then reduce to finding the maximum of n_2 occurrences of Load 2, adding the effect of this maximum to the effect of Load 1, then taking the maximum of n_1 occurrences of the combined effect of x_1 and the n_2 maximum of Load 2. This approach assumes that x_1 and x_2 have constant intensities during the duration of one of their occurrences. Notice that x_1 or x_2 could possibly have magnitudes equal to zero. If the intensities of x_1 and x_2 are random variables with known probability functions, then $X_{\max,T}$ can be calculated as described below.

The cumulative distribution of any load Y can be represented as $F_Y(Y^*)$. $F_Y(Y^*)$ gives the probability that the variable Y takes a value less than or equal to Y^*. Most load combination studies assume that the load intensities are independent from one occurrence to the other. In this case, the cumulative distribution of the maximum of m events that occur in a time period T can be calculated from the probability distribution of one event by

$$F_{Y_{\max},m}(Y^*) = F_Y(Y^*)^m \tag{2.43}$$

where m is the number of times the load Y occurs in the time period T.

Equation 2.43 is obtained by realizing that the probability that the maximum value of m occurrences of load Y is less than or equal to Y^* if the first occurrence is less than or equal to Y^*, *and* the second occurrence is less than or equal to Y^*, *and* the third occurrence is less than or equal to Y^*, and so forth. This is repeated m times, which leads to the exponent m in the right-hand-side term of Equation 2.43.

The cumulative probability distribution of $x_{2\max,t1}$ is found by plugging the probability distribution of x_2 into the right-hand side of Equation 2.43 with $m = n_2$. Then, knowing the distribution of x_1, combine the distribution of x_1 and $x_{2\max,t1}$ to find the distribution of their sum. Finally, use this latter distribution in the right-hand side of Equation 2.43 with Y representing the combined effect and $m = n_1$ to find the probability distribution of $X_{\max,T}$. Given the statistics of the resistance and the probability distribution of $X_{\max,T}$, which is the maximum load effect in the lifetime of the structure, a Level II reliability program or a Monte Carlo simulation can be used to find the reliability index β. The reliability index will provide a measure of the level of safety of a bridge member under the combined load effect, as was described in Section 2.2.

This approach, which assumes independence between the different load occurrences, has been widely used in many previous efforts on the calibration of the load factors for combined load effects. However, in many practical cases, even when the intensities of the extreme load events are independent, the random effects of these loads on the structure are not independent. For example, although the wind velocities from different windstorms may be considered independent, the maximum moments produced in the piers of bridges as a result of these winds will be functions of modeling variables such as pressure coefficients and other statistical uncertainties that are not independent from storm to storm. In this case, Equations 2.42 and 2.43 have to be modified to account for the correlation between the intensities of all m possible occurrences. For example, let us assume that the load effect x_1 of Equation 2.42 is the product of a time-dependent random load intensity z_1 and a time-independent analysis variable c_1 where c_1 describes how the load intensity is converted to a load effect. Similarly, the load effect x_2 is the product of a time-dependent random load intensity z_2 and a time-independent analysis variable c_2 that describes how the load intensity is converted to a load effect. In this case, Equation 2.42 can be expanded as

$$X_{max,T} = \max_{n_1}\left[c_1 z_1 + \max_{n_2}[c_2 z_2]\right] \tag{2.44}$$

where c_1 and c_2 are time-independent random variables that do not change with n_1 and n_2 while the load intensities z_1 and z_2 are time-dependent variables that increase as the return periods represented by the number of occurrences n_1 and n_2 increase.

The randomness in c_1 and c_2 can be accounted for by using conditional probability functions; that is, Equation 2.43 can be used to determine the distribution function of $X_{max,T}$ for pre-set values of the analysis modeling factors c_1 and c_2, which are assumed to be constant. Then, knowing the probability that c_1 and c_2 are equal to the pre-set values, the final unconditional probability of $X_{max,T}$ is obtained by summing together the products of the probability associated with each value of the modeling factor and the outcome of Equation 2.43. This can be represented by the following equation:

$$\Pr[X_{max,T} \leq X^*] = F_{X_{max,T}}(X^*) \\ = \iint F_{X_{max,T}}(X^* \mid c_1, c_2) f_{c1}(C_1) f_{c2}(C_2) dc_1 dc_2 \tag{2.45}$$

where $F_{X_{max,T}}(X^*|c_1,c_2)$ is the probability function of $X_{max,T}$ conditional on particular values of c_1 and c_2 and f_{c1} and f_{c2} are the probability density functions of the analysis variables c_1 and c_2.

The model described above based on Figure 2.14 assumes that the return period T can be exactly divided into n_1 occurrences of the Load 1 where the time duration of each occurrence is t_1 such that $T = t_1 x n_1$. Modifications on the classical model can be made to describe situations when no loads are on. For example, these can be accounted for by using a cumulative probability distribution function F with high probability values for $X = 0$. Also, if the time duration

of the loads is short, the Ferry Borges–Castanheta model can be represented as shown in Figure 2.15, in which the basic time interval t_1 represents the recurrence interval of Load 1. For example, if a load is known to occur on the average once every year, then $t_1 = 1$ year is used, and n_1 the number of repetitions in a 75-year return period will be $n_1 = 75$. Using the same logic, n_2 becomes the average number of occurrences of Load 2 in the time period, t_1. For example, if on the average there are 200 wind storms in 1 year, then $n_2 = 200$ when $t_1 = 1$ year.

2.5.1 Scour as a Special Case

The combination of the effect of scour with the other extreme events does not follow the assumptions of the models described above. This is because scour does not produce a load effect on bridge structures but changes the geometry of the bridge pier so that the effects of the other loads are amplified. This phenomenon is illustrated in Figure 2.16.

For example, let us assume that a bridge pier has a height equal to h. When the pier is subjected to a lateral force of magnitude P, the moment at the base of the pier is originally equal to Ph. The random variable P could be the force caused by wind pressure or may describe the impact force caused by the collision of a vessel with the pier. If the foundation of the pier has been subjected to a random scour depth equal to y_{scour}, then, after the occurrence of scour, the moment at the base of the pier will be equal to

$$M_{scour} = P(h + y_{scour}). \tag{2.46}$$

Thus, the combined effect of ship collision (or wind load) and scour is multiplicative, not additive as described above. The probability of failure will then be equal to the probability that M_{scour} is greater than the moment capacity of the pier or the moment capacity of the foundation. If the foundation is very deep, then the moment capacity will remain the same as it was before scour. In this case, the moment arm of the load, P, changes as the point of fixity in the foundation is shifted downward. If the foundation is not deep, the ability of the foundation to resist overtipping may decrease because of the occurrence of scour. In this case, the capacity of the foundation will need to be calculated based on the remaining foundation depth and the type of soil. Another mode of failure is the loss of serviceability caused by the lateral deflection of the pier-superstructure system. That is, the foundation may not fail, but the flexibility of the pier-foundation system after scour may produce large deformations in the structure, rendering it unfit for use.

A model to obtain the probability distribution of the maximum scour depth, y_{scour}, for different return periods is presented in Section 2.4.4. Section 2.4 also presents models to obtain the probability distribution of the maximum force effect,

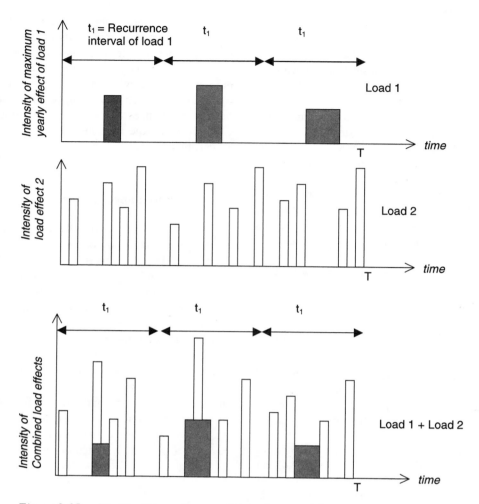

Figure 2.15.   Modified Ferry Borges–Castanheta model.

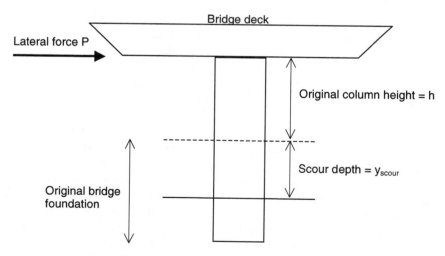

Figure 2.16.   Representation of combination of scour and other loads.

P, as a function of the return period for the extreme load events of interest to this study. The maximum moment effect caused by the combination of scour and the other extreme events can then be calculated using the expression for M_{scour} as is shown in Equation 2.46.

It is noted that the presence of scour would alter the natural period of the bridge system, rendering it more flexible. The additional flexibility induced by the presence of scour may sometimes reduce the inertial forces produced by dynamic vibrations, such as those caused by earthquakes, rendering the structure less vulnerable to these threats. Hence, the presence of scour may under certain circumstances be beneficial and improve the reliability of the structure. Therefore, it is critical to check the safety of bridges with and without scour to ensure that the worst conditions are accounted for.

2.6 CHAPTER CONCLUSIONS

This chapter described the models used in this study to perform the reliability analysis of highway bridges subjected to extreme events and the combination of extreme events. As a first step, the statistical information necessary to perform the reliability analysis has been assembled from the literature and, when possible, as used during previous code development efforts. The modified Ferry Borges–Castanheta model used in this study to perform the reliability analysis for the combination of load events has also been described. Examples illustrating the use of the proposed models are provided in the appendixes. The results are summarized and used to propose a set of load factors for the combination of extreme events, as will be described in Chapter 3.

CHAPTER 3

CALIBRATION OF LOAD FACTORS FOR COMBINATIONS OF EXTREME EVENTS

This chapter describes the procedure followed during the reliability analysis of bridges subjected to extreme events and combinations of events. The calibration of appropriate load factors is also described. Following the research approach described in Chapter 1, the first step of the process requires the calculation of the reliability index, β, for a set of typical bridge configurations designed to satisfy current AASHTO LRFD specifications. Because the AASHTO LRFD specifications are primarily concerned with medium- to short-span bridge spans, the basic configuration used in this chapter consists of a three-span bridge with span lengths of 18 m/30 m/18 m (60 ft/100 ft/60 ft). The bridge is assumed to be supported by either single-column bents in which each bent's column is 1.8 m (6 ft) in diameter or two-column bents in which each bent consists of two 1.1-m (3.5-ft) diameter columns at 8 m (26 ft) center to center. The bridge geometry and structural properties for these cases are further described in Section 3.1. The reliability analysis of these two basic bridge configurations is first performed for each of the pertinent extreme hazards individually. The results of these calculations are provided in Section 3.2. Section 3.3 gives the results of the reliability calculations for the combination of events and also describes the load factor calibration process. Section 3.4 gives a summary of the load combination factors. These are also presented in a an AASHTO specifications format in Appendix A. Additional details of the calculations described in this chapter are provided in Appendixes C and D. Some of the models and assumptions made during the calculations are based on the analyses and models presented in Appendixes B, E, F, H, and I.

3.1 DESCRIPTION OF BASIC BRIDGE CONFIGURATIONS AND STRUCTURAL PROPERTIES

The basic bridge configuration used in this study consists of a three-span 66-m (220-ft) bridge having the profile shown in Figure 3.1. The bridge is also assumed to span over a small river that may produce local scour around the bridge columns. The geometric properties of the columns for each of the bents considered are shown in Figure 3.2.

3.1.1 One-Column Bent

The first option for substructure configuration consists of a bridge with two bents, each formed by a single 1.8-m (6-ft) diameter concrete column with concrete strength, f'_c, of 28 MPa (4,000 psi). The cap beam is 1.5 to 2.1 m (5 ft to 7 ft) deep carrying 6 Type-6 AASHTO girders and a 0.25-m (10-in.) deck slab plus wearing surface. The deck is 12-m (40-ft) wide with a 0.9-m (3-ft) curb on each side. The required column capacity is calculated to satisfy each of the pertinent loads using the current AASHTO LRFD specifications. Specifically, the required moment capacity is calculated for the one-column bent when subjected to the effects of gravity loads, wind loads, earthquakes, and vessel collision forces. The required foundation depth is determined to satisfy the requirements for each load type and also for scour. In addition, the column axial capacity and foundation bearing capacity are estimated to satisfy the requirements for applied gravity loads. The applied loads and load factors used in the calculations performed in this section to estimate the structural properties of the basic configuration are those specified by the AASHTO LRFD. A short description of the procedure is provided below for each load and limit state analyzed. More details are provided in Appendixes C and D.

Gravity Loads

The weight applied on each bent is calculated as follows:

- Superstructure weight per span length = 156 kN/m (10.7 kip/ft);
- Weight of cap beam = 690 kN (154 kips); and
- Weight of wearing surface and utilities = 36 kN/m (2.5 kip/ft).

The weight of the column above the soil level is 390 kN (88 kips). The analysis of the distributed weights will produce a dead weight reaction at the top of the column equal to 5.4 MN (1,209 kips) from the superstructure plus 690 kN (154 kips) from the cap and 390 kN (88 kips) from the column, producing a total weight equal to 6.5 MN (1,450 kips).

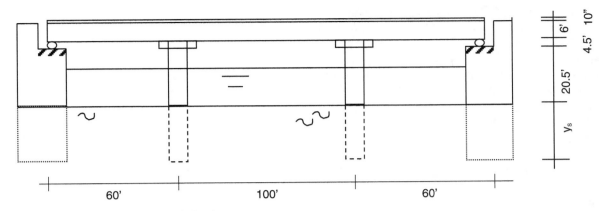

Figure 3.1.   *Profile of example bridge.*

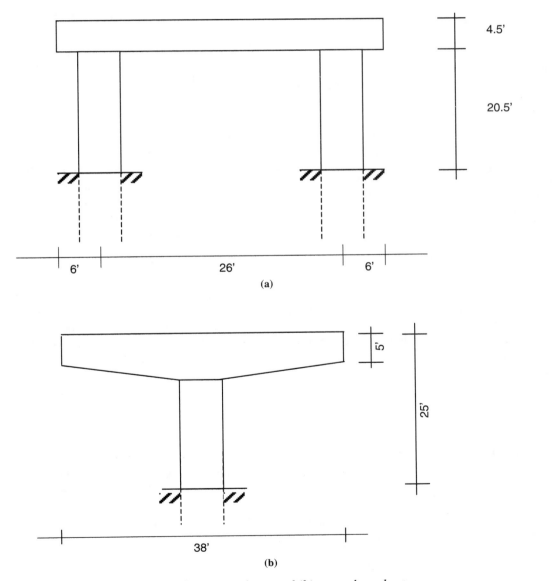

Figure 3.2.   *Configurations of (a) two-column and (b) one-column bents.*

If designed to carry gravity loads alone (live and dead loads), the column and foundations of the one-column bent should be designed to support either two lanes of HL-93 live load that produce an unfactored live load on top of the column equal to 1.9 MN (423 kips) at an eccentricity of 2 m (7 ft), or one lane of loading that is 1.1 MN (254 kips) (including a multiple presence correction factor of 1.20) at an eccentricity of 3.7 m (12 ft) from the center of the column. The free body diagram for the one-column bent subjected to gravity loads is shown in Figure 3.3 in which F_{LL} is the force caused by live load and F_{DC} is the force from the dead weight.

The design equation used for axial and bearing capacity is as follows:

$$\phi R = 1.25\,DC + 1.75\,LL \tag{3.1}$$

where

 DC = 6.5 MN (1,450 kips) = the permanent weight of components;
 LL = the live load = 1.9 MN (423 kips); and
 ϕ = the resistance factor, which for column strength is equal to 0.75 and for foundation bearing capacity is taken as 0.50; and
 R = the nominal axial capacity in kips.

The required nominal resistance capacity is obtained from Equation 3.1 for each case as listed in Table 3.1.

For bending moment capacity of the column, the following equation used is:

$$\phi R = 1.75\,LL \tag{3.2}$$

where LL is the live load moment = 4.1 kN-m (3,048 kip-ft = 254 kip × 12 ft) and ϕ is the resistance factor which for bending is equal to 0.90. In this case, R is the nominal moment capacity in kN-m.

In the case of foundation overtipping, the permanent weight of the bridge would help resist overtipping, hence the design equation used is

$$\phi R = 0.90\,DC + 1.75\,LL \tag{3.3}$$

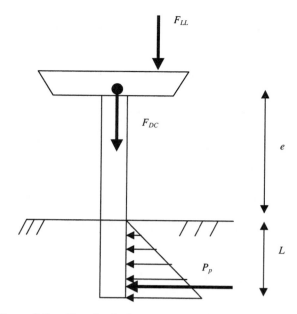

Figure 3.3. Free body diagram of one-column bent under gravity loads.

where

 DC = 5.9 kN-m (4,350 kip-ft = 1,450 kip × 3 ft) is the moment caused by the permanent weight of components about the edge of the column shaft;
 LL = the live load moment about the edge of the column
 LL = 3.1 kN-m (2,286 kip-ft = 254 kip × 9 ft); and
 ϕ = the resistance factor, which for foundation lateral resistance is taken as 0.50.

The 2.7-m (9-ft) moment arm is used as the distance between the live load and the edge of the column rather than the center, which is used when calculating the column bending moment. In this case, also R is a moment capacity in kN-m. The required foundation depth (embedded length), L, is obtained from the R value calculated from Equation 3.3, assuming a triangular distribution of soil pressure using the free body diagram shown in Figure 3.3. If P_p is calculated from Equations 2.15 and 2.16 of Chapter 2 using a soil unit weight, $\gamma = 960$ kg/m³ (60 lb/ft³); an angle of friction for sand,

TABLE 3.1 Required design capacities of one-column bent under gravity load

Hazard	Member	Limit State	Current Design Load Factors	Resistance Factor ϕ	Required Nominal Capacity
Gravity load	Column	Axial capacity	1.25 DC + 1.75 LL	0.75	3,404 kips
		Moment capacity	1.25 DC + 1.75 LL	0.90	5,908 kip-ft
	Foundation	Vertical bearing capacity	1.25 DC + 1.75 LL	0.50	5,106 kips
		Foundation depth to prevent overtipping	0.90 DC + 1.75 LL	0.50	6 ft

$\phi_s = 35°$; and a column diameter, $D = 1.8$ m (6 ft); then L is found by setting $R = P_pL/3$. Knowing R from Equation 3.3, the foundation depth, L, required to resist overtipping can be calculated as $L = 1.8$ m (6 ft).

Table 3.1 summarizes the results of the nominal design capacities for the column axial force, column bending moment, foundation bearing resistance, and foundation depth for the one-column bent.

Wind Loads

The design of the single-column bridge bent for wind load in interior regions of the United States uses a design wind speed of 145 km/h (90 mph) as stipulated in the AASHTO LRFD wind maps. The AASHTO LRFD design equation for the moment capacity of the column when the system is subjected to wind is given as follows:

$$\phi R = 1.40\, WS \tag{3.4}$$

where R is the required moment capacity of the column and WS is the applied moment. For overtipping about the base of the foundation system, the equation used is of the form

$$\phi R = 0.90\, DC + 1.40\, WS. \tag{3.5}$$

In Equation 3.5, DC is the moment produced by the permanent weight about the bottom edge of the column, and WS is the moment about the base of the column produced by the wind load on the structure.

In Equation 3.4, the maximum bending moment in the column occurs at a point located at a distance f below the soil surface. f corresponds to the point in which the lateral force P_p in the soil is equal to the applied lateral force from the wind. The free body diagram for column bending is illustrated in Figure 3.4(b). For Equation 3.5, which checks the safety of the system against overtipping about the base, the free body diagram is illustrated in Figure 3.4(a).

The forces F_1 applied on the superstructure and F_2 applied on the column because of the wind are calculated from the AASHTO LRFD wind pressure equation, which is given as follows:

$$P_D = P_B \frac{V_{DZ}^2}{25,600} \tag{3.6}$$

where P_B is base wind pressure = 0.0024 MPa for beams and V_{DZ} is the design wind velocity at design elevation Z, which is calculated from

$$V_{DZ} = 2.5V_0 \left(\frac{V_{10}}{V_B}\right)\ln\left(\frac{Z}{Z_0}\right) \tag{3.7}$$

where

V_{10} = wind velocity above ground level = 145 km/h (90 mph) in the case studied herein;
V_0 = friction velocity = 13.2 km/h (8 mph) for open country;
V_B = base wind velocity = 160 km/h (100 mph); and
Z_0 = friction length = 70 mm for open country.

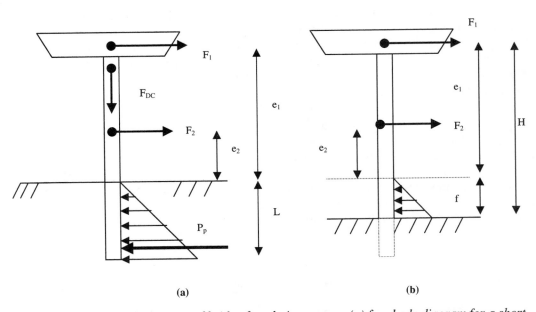

(a)   (b)

Figure 3.4.   Free body diagram of bridge foundation system: (a) free body diagram for a short pile, dominant failure mode is tipping at base; (b) diagram for a long pile, dominant failure mode is column bending at distance f below soil level.

Using a superstructure depth of 2.4 m (8 ft), the wind force transmitted from the superstructure to the column is calculated to be $F_1 = 140$ kN (31.6 kips) applied at a height $e_1 = 8.8$ m (29 ft) from the soil surface. The force applied on the substructure is $F_2 = 27$ kN (6 kips) at a height $e_2 = 3.6$ m (11.8 ft). For long columns, the point of maximum moment is at $f = 1.6$ m (5.14 ft) below the soil surface. Thus, the moment from the wind on the structure $WS = 1600$ kN (1,180 kip-ft). The corresponding nominal design bending moment of 2500 kN-m (1,835 kip-ft) is obtained from Equation 3.4 using a resistance factor $\phi = 0.90$.

The results of Equation 3.5 show that the column is able to resist overtipping simply because of the counteracting effects of the permanent weight. Thus, the foundation depth is controlled by effects other than the wind load (e.g., the effect of live load as shown in the previous section). Table 3.2 summarizes the wind load design requirements.

Earthquakes

When performing a dynamic analysis in the transverse direction and assuming that the point of fixity for lateral deformation is 5.5 m (18 ft) below soil level, the effective weight of the column above the point of fixity is equal to 747 kN (168 kips). In this latter case, the center of mass is 8 m (27 ft) above soil level (= 45 ft from the point of fixity). Following common practice in earthquake engineering and assuming a tributary length of 28 m (91.75 ft = 50% of the distance to other bent and 70% of distance to abutment), the total weight from the superstructure and wearing surface applied on one bent adds up to 5360 kN (1,205 kips). Thus, the total weight on one bent is equal to 6800 kN (1,527 kips). This assumes that the lateral connection of the superstructure to the abutments will not break because of the earthquake-induced lateral motions. Notice that the total weight is slightly higher than for the gravity load because of the inclusion of the weight of the pile up to the point of fixity. The center of mass is calculated to be at 8.2 m (27 ft) above the soil surface.

The inertial forces applied on the bent because of the earthquake accelerations may be represented as shown in Figure 3.5. The point of fixity at a distance L_e below the soil surface can be calculated as a function of the soil elastic modulus, E_s, the foundation depth, L, and the stiffness of the column represented by $E_p I_p$ where E_p is the modulus of the concrete pile and I_p is the moment of inertia of the concrete pile. The foundation is assumed to consist of a drilled pile shaft (pile extension) that

extends 15 m (50 ft) into the soil. The soil is assumed to have an elastic modulus $E_s = 69$ MPa (10,000 psi) corresponding to moderately stiff sand. The point of fixity of the floating foundation can be calculated using the relationship provided by Poulos and Davis (1980) given as follows:

$$\left(\frac{L_e}{L}\right)^3 + 1.5\frac{e}{L}\left(\frac{L_e}{L}\right)^2 = 3K_R\left(I_{\rho H} + \frac{e}{L}I_{\rho M}\right) \qquad (3.8)$$

where

  L_e = the effective depth of the foundation (distance from ground level to point of fixity);
  L = the actual depth;
  e = the clear distance of the column above ground level;
  K_R = the pile flexibility factor, which gives the relative stiffness of the pile and soil;
  $I_{\rho H}$ = the influence coefficient for lateral force; and
  $I_{\rho M}$ = the influence coefficient for moment.

The pile flexibility factor is given as follows:

$$K_R = \frac{E_p I_p}{E_s L^4}. \qquad (3.9)$$

If the pile is made of 28 MPa (4,000 psi) concrete, then $E_p = 25$ GPa $(3,600$ ksi $= 57\sqrt{4,000})$ and the diameter of the column being $D = 1.8$ m (6 ft) result in a moment of inertia $I_p = 0.55$ m^4 (63.62 ft^4 = $\pi r^4/4$). Thus, for a pile length of 15 m (50 ft), the pile flexibility becomes $K_R = 0.0037$. The charts provided by Poulos and Davis (1980) show that for $K_R = 0.0037$ the influence coefficients $I_{\rho H}$ and $I_{\rho M}$ are respectively on the order of 5 and 15. The root of Equation 3.8 produces a ratio $L_e/L = 0.35$, resulting in an effective depth, L_e, of 18 ft below ground surface. Thus, the effective total column height until the point of fixity becomes $H = 13.7$ m (45 ft = 27 ft + 18 ft).

The natural period of the column bent can be calculated from

$$T = 2\pi\sqrt{\frac{M}{K}} \qquad (3.10)$$

where the column stiffness is given as $K = 3E_p I_p/H^3$ and the mass M is the equal to weight 6.8 MN (1,527 kips) divided by g, the acceleration due to gravity. Given the material and geometric properties, the natural period of the column is found to be $T = 1.31$ sec.

TABLE 3.2 Required design capacities of one-column bent under wind load

Hazard	Member	Limit State	Current Design Load Factors	Resistance Factor ϕ	Required Nominal Capacity
Wind load	Column	Moment capacity	1.40 *WS*	0.90	1,835 kip-ft
	Foundation	Overtipping	0.90 *DC* + 1.40 *WS*	0.50	0 ft

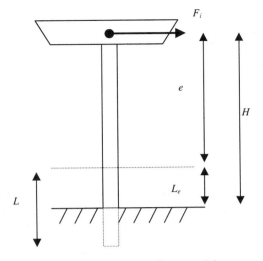

Figure 3.5. Dynamic analysis model.

The natural period of the system is used along with Equations 2.17 through 2.19 and Table 2.4 of Chapter 2 to find the spectral accelerations, S_a, for different bridge site locations. The soil is assumed to be of type D. The spectral accelerations calculated for the 2,500-year return period (2% probability of exceedance in 50 years) are shown in Table 3.3. The equivalent inertial forces, F_i, for nonlinear columns are obtained using Equation 3.11:

$$F_i = \frac{S_a W}{R_m} \qquad (3.11)$$

where

S_a = the spectral acceleration as a function of the acceleration due to gravity, g;
W = the weight of the structure; and
R_m = the response modification factor.

Assuming a response modification factor $R_m = 1.50$ as specified in the proposed AASHTO LRFD earthquake design specifications in *NCHRP Report 472* (ATC and MCEER, 2002), the equivalent inertial forces accounting for column nonlinearity are calculated as shown in Table 3.3. Using the free body diagrams of Figure 3.4(b) with $F_1 = F_i$ and $F_2 = 0$, the required moment capacities for the bridge column for different earthquake sites are given as shown in Table 3.3.

Table 3.3 also shows the required foundation depth, L, calculated using the free body diagram of Figure 3.4(a). In this case, the R_m value used is $R_m = 1.0$ because rigid body overtipping is assumed to occur in short columns where the maximum moment capacity has not been reached yet (i.e., when the column's bending moment is still in the linear elastic range). For these cases, the required foundation depth, L, is given in the last column of Table 3.3.

Note that following common practice, the stiffness and the natural period of the system are calculated using Equations 3.8 and 3.9, which are based on the elastic behavior of the system. The design uses Equation 3.11 and the free body diagram of Figure 3.4, which are based on ultimate capacity.

Scour

It is herein assumed that the bridge is constructed over a 67-m (220-ft) wide rectangular river channel as shown in Figure 3.1. To obtain realistic results for the effect of scour, different possible discharge rate data from different small rivers are used, and design scour depths are calculated for each of these river discharge rates. The rivers chosen for this analysis consist of the following: (1) Schohaire Creek in upstate New York, (2) Mohawk River in upstate New York, (3) Cuyahoga River in northern Ohio, (4) Rocky River in Ohio, and (5) Sandusky River in Ohio. Data on the peak annual discharge rates for each of the five rivers were obtained from the USGS website. Lognormal probability plots and Kolmogorov–Smirnov (K-S) goodness-of-fit tests showed that the peak annual discharge rate, Q, for all five rivers can be reasonably well modeled by lognormal probability distributions. The mean of the log Q and its standard deviation were calculated using a maximum likelihood estimator. These data are provided in Table 2.8 of Chapter 2. Assuming a slope of $S_0 = 0.2\%$ and a Manning roughness coefficient of $n = 0.025$, the flow depth for the 100-year flood, y_{max}, can be calculated along with the corresponding flow velocity, V, for each of the five rivers, using Equations 2.22 through 2.28. Assuming a 67-m (220-ft) wide rectangular channel and a pier diameter of $D = 1.8$ m (6 ft), the maximum design scour depth for the one-column bent bridge for each river data is obtained as shown in Table 3.4.

To avoid failure caused by scour, the foundation depth should be designed for a length, L, greater than or equal to the design scour depth, Y_{max}, which is given in Table 3.4.

TABLE 3.3 Earthquake design requirements for five sites

Site	Spectral Acceleration, S_a	Equivalent Inertial Force, F_i	Required Moment Capacity, M_{cap}	Required Foundation Depth, L
San Francisco	1.15 g	1,171 kips	55,950 kip-ft	86 ft
Seattle	0.64 g	652 kips	28,320 kip-ft	66 ft
Memphis	0.50 g	510 kips	21,050 kip-ft	59 ft
New York	0.17 g	173 kips	6,500 kip-ft	37 ft
St. Paul	0.05 g	51 kips	1,670 kip-ft	20 ft

56

TABLE 3.4 Design scour depth for five rivers

River	Q 100-year (ft³/sec)	V (ft/sec)	Y_0 - flood depth (ft)	Y_{max} – design scour depth (ft), one-column bent
Schohaire	78,146	17.81	20.56	17.34
Mohawk	32,747	12.87	11.78	13.99
Sandusky	36,103	13.35	12.52	14.33
Cuyahoga	19,299	10.50	8.45	12.26
Rocky	19,693	10.58	8.56	12.32

Vessel Collision

The basic bridge configuration shown in Figure 3.1 does not allow for the passage of vessels under the bridge span. Hence, a different bridge configuration is assumed to study the vessel collision problem. For the purposes of this study, the bridge configuration selected is similar to that of the Interstate 40 (I-40) bridge over the Mississippi River in Memphis, which is fully described in Appendix D. The bridge pier and water channel configurations are represented as shown in Figure 3.6. USACE provided data on the types of barges that travel the Mississippi River near Memphis along with their weights. The data are used along with the model of the AASHTO guide specifications (1991) following the recommendations proposed by Whitney et al. (1996) to find the required design vessel impact force that would produce a nominal annual failure rate of $AF = 0.001$.

The calculations use Equation 2.32 through 2.38, which are given in Chapter 2. According to the analysis summarized in Table 3.5 the design impact force should be $H_{CV} = 35$

MN (7,900 kips). This design impact force is assumed to be applied at 4.9 m (16 ft) above soil level on a column that has a diameter D = 6.1 m (20 ft). The free body diagram for designing the column and foundation capacities is similar to that depicted in Figure 3.4, with $F_1 = 0.0$, $F_2 = H_{CV}$, and $e_2 = 4.9$ m (16 ft). Given the very large permanent weight that is applied on the bent from the large superstructure needed to span the river channel, the possibility of overtipping because of vessel collision is negligibly small. Hence, only two limit states are investigated for this case: (1) failure of the column in shear, and (2) failure of the column in bending moment.

The design for shear is checked using an equation of the form

$$\phi R = 1.00\, CV \tag{3.12}$$

where CV is the vessel collision force $CV = H_{CV} = 35$ MN (7,900 kips) and is the resistance factor for shear, which according to the AASHTO LRFD should be equal to 0.90 for normal density concrete. This would result in a required shear

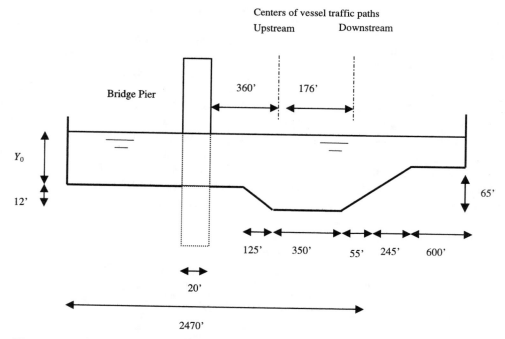

Figure 3.6. Profile of river channel for vessel collisions.

TABLE 3.5   Details of calculations of annual failure rates for vessel collisions by flotilla type

Type[a]	Frequency	Number of		Mean			Distances (m)		Normalized		PG[d]	PA[d]	PC[d]	N*PA*PG*PC
Down-bound		columns	rows	weight(MN)	LOA(m)[b]	Bm[c]	d1[c]	d2[c]	d1	d2				
AB	91	4	1.9	22.92	18.74	9.45	152.56	176.62	2.04	2.36	0.01	1.77E-04	0.00	0.00
BC	126	5.93	2.05	138.44	44.52	16.11	145.03	184.15	0.55	0.70	0.05	1.77E-04	0.00	0.00
BD	1	5.11	2.93	128.28	45.35	16.49	137.39	191.80	0.59	0.83	0.07	1.77E-04	0.00	0.00
CC	79	5.11	2.93	92.63	54.70	12.24	143.62	185.57	0.51	0.66	0.05	1.77E-04	0.00	0.00
DC	35610	5.38	3.21	116.19	59.44	10.67	157.73	171.45	0.52	0.56	0.02	1.77E-04	0.00	0.00
EA	6	5	1	119.22	60.96	7.62	156.90	172.29	0.51	0.57	0.02	1.77E-04	0.00	0.00
EB	6	5	1	124.82	60.96	9.30	156.04	172.42	0.51	0.57	0.02	1.77E-04	0.00	0.00
EC	33026	5.38	3.3	129.44	61.53	10.81	143.71	185.47	0.43	0.56	0.04	1.77E-04	0.00	0.00
FC	1043	4.5	3	209.20	79.81	15.96	137.60	191.58	0.38	0.53	0.05	1.77E-04	0.00	0.00
GC	648	4	2.8	195.75	90.01	16.17	138.91	190.27	0.39	0.53	0.05	1.77E-04	0.00	0.00
GD	19	3.23	2.3	173.97	90.71	16.49	142.58	186.60	0.49	0.64	0.05	1.77E-04	0.00	0.00
HC	17	3.23	2.3	166.98	96.53	16.17	142.95	186.24	0.46	0.60	0.05	1.77E-04	0.00	0.00
Up-bound														
AB	219	4	1.9	22.60	18.73	9.42	97.73	121.73	1.30	1.63	0.04	1.77E-04		
AC	78	4	1.9	72.28	30.30	16.46	91.04	128.41	0.75	1.06	0.08	1.77E-04		
BB	4	4	1.9	9.97	35.05	8.84	98.28	121.17	0.70	0.86	0.05	1.77E-04		
BC	760	5.93	2.05	147.40	44.40	15.99	90.29	129.16	0.34	0.49	0.05	1.77E-04	1.97E-02	1.43E-04
BD	151	5.11	2.93	138.53	45.35	16.49	82.52	136.93	0.36	0.59	0.08	1.77E-04	1.15E-02	2.57E-05
CC	396	5.11	2.93	159.39	53.79	16.07	83.14	136.32	0.30	0.50	0.07	1.77E-04	1.69E-02	8.45E-05
DC	14698	5.38	3.21	119.36	59.44	10.74	89.43	130.02	0.28	0.41	0.05	1.77E-04		
EB	1	5	1	116.89	60.96	9.30	102.03	117.42	0.33	0.39	0.02	1.77E-04		
EC	12097	5.38	3.3	131.09	61.62	11.05	88.45	131.01	0.27	0.40	0.05	1.77E-04	2.90E-03	3.01E-04
ED	1	4.5	3	68.03	73.46	17.68	80.16	139.29	0.24	0.42	0.07	1.77E-04		
FC	697	4.5	3	193.33	81.43	16.02	82.66	136.80	0.23	0.37	0.06	1.77E-04	1.96E-02	1.36E-04
FD	2	4.5	3	125.15	81.38	17.68	80.16	139.29	0.22	0.38	0.06	1.77E-04	3.20E-03	6.97E-08
GC	2028	4	2.8	197.23	90.03	16.25	83.93	135.52	0.23	0.38	0.05	1.77E-04	1.58E-02	3.10E-04
GD	158	3.23	2.3	145.12	90.71	16.49	87.72	131.74	0.30	0.45	0.06	1.77E-04		
HC	73	3.23	2.3	145.11	96.55	15.67	88.66	130.79	0.28	0.42	0.05	1.77E-04		
HD	17	2	1	103.57	117.98	21.76	95.80	123.65	0.41	0.52	0.04	1.77E-04		

NOTES:
(a) The definitions of barge flotilla types are provided in Appendix E (on *CRP-CD-30*).
(b) LOA = length overall of flotilla.
(c) These terms are defined in Figure 2.12, shown as x_1 and x_2.
(d) See Section 2.4.5 for definitions of the probabilities *PA*, *PG*, and *PC*.

column capacity equal to V_{cap} = 39 MN (8,780 kips). The point of fixity is found to be at f = 10.5 m (34.5 ft) below the soil level. Using the free body diagram of Figure 3.4(b) and a moment resistance factor, ϕ, also equal to 0.90, the required column moment capacity is found equal to M_{cap} = 464 MN-m (342,333 kip-ft). The design requirements for the column subjected to vessel collision are summarized in Table 3.6.

3.1.2 Two-Column Bents

The second design option is for the same bridge superstructure described in Figure 3.1, but where the substructure consists of two bents, each formed by two 1.1-m (3.5-ft) diameter columns spaced at 7.9 m (26 ft) center to center as described in Figure 3.2(a). The clear column height is 6.3 m (20.5 ft) connected by a 1.4-m (4.5-ft) cap carrying six Type-6 AASHTO girders and a 10-in. deck slab plus wearing surface. The bent carries the same 12.2-m (40-ft) wide deck with a 0.91-m (3-ft) curb on each side as that of the one-column bent. The required column capacities are calculated to satisfy each of the pertinent loads using the current AASHTO LRFD specifications. Specifically, the required moment capacity is calculated for the two-column bent when subjected to the effects of wind loads, earthquakes, and vessel collision forces. The analysis assumes that the bent cap is very stiff compared

TABLE 3.6   Required design capacities of one-column bent under vessel collision forces

Hazard	Member	Limit State	Current Design Load Factors	Resistance Factor ϕ	Required Nominal Capacity
Vessel Collision	Column	Moment capacity	1.00 *VC*	0.90	342,333 kip-ft
	Column	Shearing capacity	1.00 *VC*	0.90	8,780 kips

with the columns. Thus, vertical loads will produce negligible moments in the columns, and only the lateral loads will produce moments. Because of the presence of the two columns, no overtipping is possible, although foundation depth is controlled by the scour requirements. In addition, the column axial capacity and foundation bearing capacity are determined to satisfy the requirements for applied gravity loads. The applied loads and load factors used in the calculations performed in this section are those specified by the AASHTO LRFD. A short description of the procedure is provided below for each load and limit state analyzed. More details are provided in Appendix C.

Gravity Loads

The weight applied on each bent is calculated as follows:

- Superstructure weight = 156 kN/m (10.67 kip/ft),
- Weight of cap beam = 480 kN (109 kips), and
- Weight of wearing surface = 35.9 kN/m (2.46 kip/ft).

The weight of the each column above the soil level is 138 kN (31 kips). The analysis of the distributed weights will produce a dead weight reaction at the top of the bent equal to 5400 kN (1,209 kips) from the superstructure plus 480 kN (109 kips) from the cap. These will result in 2900 N (659 kips) for each column for a total of 3000 kN (690 kips) per column.

If designed to carry vertical loads alone, each column of the two-column bent should be designed to support two lanes of HL-93 live load that produce an unfactored live load on top of the column equal to 188 kN (423 kips) at an eccentricity of 1.8 m (6 ft). For one lane of traffic, the live load is 1100 kN (254 kips) at an eccentricity of 0.3 m (1 ft). The free body diagram is shown in Figure 3.7 with the lateral forces $F_1 = F_2 = 0$. In Figure 3.7, e_3 denotes the eccentricity of the applied live load, F_{LL}, relative to the center of a column. Assuming a rigid cap beam compared with the column stiffness, the applied moment on the columns caused by vertical loads is small, and the columns need to resist mostly vertical loads. Using design Equation 3.1 results in a required nominal column axial capacity and a nominal foundation bearing capacity, as are shown in Table 3.7.

Wind Loads

The design of the two-column bridge bent for wind load in interior regions of the United States uses a design wind speed

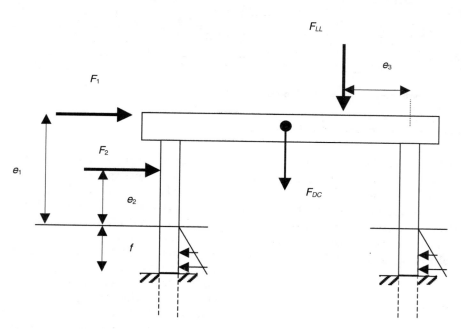

Figure 3.7. Basic free body diagram of two-column bent.

TABLE 3.7   Required design capacities of two-column bent under gravity load

Hazard	Member	Design Parameter	Current Design Load Factors	Resistance Factor ϕ	Required Nominal Capacity
Gravity load	Column	Axial capacity	1.25 *DC* + 1.75 *LL*	0.75	1,909 kips
	Foundation	Vertical bearing capacity	1.25 *DC* + 1.75 *LL*	0.50	2,864 kips

of 145 km/h (90 mph) as stipulated in the AASHTO LRFD wind maps. The free body diagram for the wind analysis problem is as shown in Figure 3.7 with $F_{LL} = 0$, $F_1 = 140$ kN (31.6 kips) at $e_1 = 8.8$ m (29 ft), and $F_2 = 18$ kN (3.95 kips) at $e_2 = 4$ m (13 ft). The design column bending moment capacity to resist the applied lateral caused by wind and allowing for a resistance factor of 0.90 and a wind load factor of 1.40 is equal to 600 kN-m (440 kip-ft) for each column as shown in Table 3.8. The simplified analysis model assumes fixity at the top because of the presence of a stiff column cap and at the base at a distance $f = 1.4$ m (4.6 ft) below the soil level.

Earthquakes

When performing a dynamic analysis in the transverse direction and assuming that the point of fixity for lateral deformation is 3 m (10 ft) below soil level, the effective weight of the column above the point of fixity is equal to 200 kN (45 kips). In this case, the center of mass is 8 m (27 ft) above soil level (= 11.3 m or 37 ft from the point of fixity). Following common practice in earthquake engineering and assuming a tributary length of 28 m (91.75 ft = 50% of the distance to other bent and 70% of distance to abutment), the total weight of superstructure and wearing surface add up to 5360 kN (1,205 kips) for one bent. This will result in a total weight on one column equal to 2.9 MN (650 kips).

As a first step it is assumed that the foundation consists of a drilled pile shaft (pile extension) that extends 15 m (50 ft) into the soil. A sensitivity analysis is performed at a later stage to check other foundation depths. The soil is assumed to have an elastic modulus $E_s = 10,000$ psi corresponding to moderately stiff sand. As explained in Appendix C, the point of fixity of the floating foundation can be calculated using the relationship provided by Poulos and Davis (1980) to produce a point of fixity $L_e = 3$ m (10 ft) below soil level. Given a clear height of the column $e = 6.2$ m (20.5 ft), the effective total column height until the point of fixity becomes 9.3 m (30.5 ft = 20.5 ft + 10 ft). For transverse seismic motion, the bent is

taken as fixed at the base of the effective pile depth and also fixed on the top of the column because of the presence of the stiff column cap. Thus, the natural period of the bent system, T, is 0.72 sec. The natural period is used to find the inertial forces and the required moment capacities for different earthquake sites using the same method described in Section 3.1.1 for bridges with one-column bent as provided in Equation 3.10, but with $K = 12E_p I_p/L^3$. Given the natural period, T, the spectral accelerations are calculated from Equation 2.17 through 2.19 of Chapter 2. The inertial forces are calculated using Equation 3.11 with a response modification factor $R_m = 1.5$. The spectral accelerations along with the inertial forces are listed in Table 3.9 for earthquake data from five different sites. The free body diagram described in Figure 3.7 is used with F_1 equal to the inertial force while all other forces are set equal to zero. The required column moment capacities are calculated as shown in Table 3.9 using a resistance factor, $\phi = 0.90$.

Scour

The analysis of the two-column bent for scour follows the same exact method used for the one-column bent. The only difference is the diameter of the column where $D = 1.1$ m (3.5 ft) is used for the two-column bent rather than the value $D = 1.8$ m (6 ft), which is used for the one-column bent. The required foundation depths for different river flood data are shown in Table 3.10. The required foundation depth, L, for each river should then be equal to or greater than the values of y_{max} given in Table 3.10.

Vessel Collision

The vessel collision design force of $H_{cv} = 35$ MN (7,900 kips) for the two-column bent is calculated following the same exact procedure described above for the single-column bent. The free body diagram for determining the maximum moment in the column is as shown in Figure 3.7, with $F_2 = H_{CV}$ at $e_2 = 4.9$ m (16 ft) and $F_1 = F_{LL} = 0$. In the case of the

TABLE 3.8   Required design capacities of two-column bent for wind load

Hazard	Member	Design Parameter	Current Design Load Factors	Resistance Factor ϕ	Required Nominal Capacity
Wind load	Column	Moment capacity	1.40 *WS*	0.90	440 kip-ft

TABLE 3.9   Earthquake design requirements for two-column bent using earthquake data from five sites

Site	S_a Spectral Acceleration	F_i Equivalent Inertial Force	M_{cap}, Required Moment Capacity
San Francisco	1.81 g	784 kips	13,500 kip-ft
Seattle	1.19 g	516 kips	8,370 kip-ft
Memphis	0.92 g	399 kips	6,260 kip-ft
New York	0.32 g	139 kips	1,970 kip-ft
St. Paul	0.09 g	39 kips	514 kip-ft

TABLE 3.10 Design scour depth for five rivers

River	Q 100-year (ft^3/sec)	V (ft/sec)	Y_0 flood depth (ft)	Y_{max} – design scour depth (ft), two-column bent
Schohaire	78146	17.81	20.56	12.31
Mohawk	32747	12.87	11.78	9.93
Sandusky	36103	13.35	12.52	10.17
Cuyahoga	19299	10.50	8.45	8.70
Rocky	19693	10.58	8.56	8.75

vessel collision analysis, the column width is taken as 6.1 m (20 ft) and the clear height of the bent is 46 m (150 ft), which is the same height as the I-40 bridge over the Mississippi River described in Appendix D. Because the clear column height of 46 m (150 ft) is so much higher than the point of application of the force, which is at 4.9 m (16 ft) above the soil level, most of the applied load remains in the lower portion of the impacted column. The maximum shearing force in the lower part of the column is calculated to be 0.95 times the applied force. Hence, the design shearing capacity should be 37 MN (8,340 kips) using a resistance factor $\phi = 0.90$. The point of maximum moment is approximated to be at 10 m (33 ft) below the soil surface. Hence, using a resistance factor $\phi = 0.90$, the required moment capacity at the base will be 430 MN-m (317,000 kip-ft). The required moment and shear capacities are listed in Table 3.11

3.2 RELIABILITY ANALYSIS FOR EXTREME EVENTS

The reliability index, β, is calculated in this section for bridges designed to satisfy the current AASHTO LRFD criteria for the various extreme load events of interest to this project. The purpose is to study the level of bridge structural safety provided by the current AASHTO LRFD for each of the extreme events. Also, the reliability analysis shows how changes in the design criteria would affect the safety of bridge systems. This information will also be used to choose the target reliability indexes that will form the basis for calibrating the load factors for load combinations. The reliability analysis uses the methodology and the statistical models described in Chapter 2. In this section, the reliability analysis is performed for gravity loads consisting of live load and dead loads, for wind loads in combination with the permanent loads when applicable, for earthquakes alone, for scour, and for vessel collisions. The bridges analyzed have the basic geometries described in Figures 3.1 and 3.2. Modifications on the basic

bridge geometry are assumed in order to allow for barge traffic under the bridge and, thus, also to study the reliability under vessel collision.

The reliability calculations are performed using a Monte Carlo simulation. The random numbers were generated using the method described by Deng and Lin (2000). This method has proven to be more stable and capable of generating large numbers of independent random variables than are similar routines provided in commercially available mathematical packages. An excessively large number of independent random variables are needed in this study in which Monte Carlo simulations have to be performed for a relatively large number of random variables and to study probabilities of failure on the order of 10^{-5} to 10^{-6} (reliability index, β, up to the range of 4.0). Monte Carlo simulations are used rather than a FORM algorithm because some of the cumulative distributions identified in Chapter 2 (e.g., for earthquake intensities and vessel collision forces described in Figures 2.4 and 2.13) are given in discrete forms that render them difficult to implement in a FORM algorithm because of the discontinuities of their slopes.

3.2.1 Reliability Analysis for Gravity Loads

In this section, the reliability index implicit in the AASHTO LRFD design criteria is calculated for bridge columns subjected to the effect of live loads in combination with dead loads. Four different limit states are studied:

1. Failure of the column under axial load,
2. Failure of the foundation caused by exceedance of bearing capacity,
3. Overtipping of a column with a short pile shaft, and
4. Failure of a long pile shaft in bending.

For the analysis of the column for overtipping, the soil's contributions in resisting bending are included using the free body diagram in Figure 3.3. The failure functions for each of

TABLE 3.11 Required design capacities of two-column bent under vessel collision forces

Hazard	Member	Limit State	Current Design Load Factors	Resistance Factor ϕ	Required Nominal Capacity
Vessel Collision	Column	Moment capacity	1.00 VC	0.90	317,000 kip-ft
	Column	Shearing capacity	1.00 VC	0.90	8,340 kips

the four limit states considered for one-column bents are given as follows:

$$Z_1 = P_{\text{col}} - F_{DC} - F_{LL}, \qquad (3.13)$$

$$Z_2 = P_{\text{soil}} - F_{DC} - F_{LL}, \qquad (3.14)$$

$$Z_3 = \frac{3\gamma D K_p L^2}{2} \frac{L}{3} + F_{DC} \frac{D}{2} - F_{LL}\left(e_3 - \frac{D}{2}\right), \text{ and} \qquad (3.15)$$

$$Z_4 = M_{\text{col}} - F_{LL} e_3. \qquad (3.16)$$

For each limit state, i, failure occurs when Z_i is less than or equal to zero. P_{col} is the column's axial capacity, F_{DC} is the dead weight applied on the column, F_{LL} is the live load, P_{soil} is the soil's bearing capacity at the column's tip, γ is the specific weight of the soil, D is the column diameter, K_p gives the Rankine coefficient, L is the foundation depth, e_3 is the eccentricity of the live load relative to the center of the column, and M_{col} is the column's bending moment capacity. All the variables used in Equations 3.13 through 3.16 are considered random except for the column diameter, D, the eccentricity, e_3, and the foundation depth, L. Adjustments to Equations 3.13 and 3.14 are made when analyzing the two-column bents to find the portion of F_{LL} and F_{DC} applied on each column. The statistical models used to describe the random variables are provided in Tables 2.1, 2.2 and 2.3. Specifically, the live load F_{LL} is composed of time-dependent and time-independent random variables and can be represented as follows:

$$F_{LL} = \lambda_{LL} I_{LL} HL_{93} I_{IM} \qquad (3.17)$$

where

λ_{LL} = the live load modeling factor,
I_{LL} = the live load intensity coefficient,
HL_{93} = the AASHTO LRFD live load model, and
I_{IM} = the dynamic amplification factor.

All the terms in Equation 3.17 except for HL_{93} are random with statistical properties provided in Table 2.3 of Chapter 2. I_{LL} is a time-dependent random variable that varies as a function of the return period as described in Equations 2.11 and 2.14.

It is noted that the columns are subjected simultaneously to axial compression and bending moment. Thus, the column's load-carrying capacity is governed by a P-M (axial load versus moment) interaction curve as shown in Figure 2.2. However, for the sake of simplicity and because reliability models for the combined behavior of bridge columns are not readily available, Equations 3.13 through 3.16 treat each effect separately.

For the one-column bent, the column is first assumed to have the nominal capacities provided in Table 3.1. The columns of the two-column bent are first assumed to have the nominal capacities listed in Table 3.7. In addition, a sensitivity analysis is performed to study how the reliability index

for each limit state changes as the nominal capacities are varied from the required AASHTO LRFD values of Tables 3.1 and 3.7. During the reliability calculations, the resistance biases, COVs, and probability distribution types listed in Tables 2.1 and 2.2 of Chapter 2 are used to model the variables that contribute to resisting the failure in each limit state.

Statistical data for modeling the dead and live loads' effects are described in Section 2.4.1 of Chapter 2. The biases, COVs, and probability distributions of all the random variables that describe the effects of gravity loads are provided in Table 2.3. These values are used as input during the reliability calculations. Following the model used by Nowak (1999), the bridge is assumed to carry 1,000 individual heavy truck load events or 67 occurrences of heavy side-by-side truck events each day. This would result in 27.4×10^6 single truck events or 1.825×10^6 side-by-side events in a 75-year design period. The 75-year design period is used in this report to remain compatible with the AASHTO LRFD specifications. Equation 2.14 with $m = 27.4 \times 10^6$ or 1.825×10^6 is used to obtain the probability distribution of the maximum live load intensity in 75 years given the probability distribution for one event as listed in Table 2.3. The reliability calculations also account for the modeling factor and the dynamic amplification factors. Only the live load intensity is considered to be time-dependent. All the other variables are time-independent in the sense that they are not affected by the 75-year design period. During each cycle of the Monte Carlo simulation, the worst of the one-lane loading or the two-lane loading is used as the final load applied on the bridge structure.

The results of the reliability analysis for the four limit states considered are presented in Figures 3.8 through 3.11 for the two- and one-column bents. The figures give the reliability index, β, for each limit state as a function of the nominal capacity of the column. The abscissas are normalized such that a value of 1.0 indicates that the column is designed to exactly satisfy the requirements of the AASHTO LRFD specifications for the limit state under consideration. The required nominal design capacities for the columns under the effect of gravity loads are listed in Tables 3.1 and 3.7 for the one- and two-column bents. Thus, a value of 1.0 for the failure of the column under axial load (i.e., the first limit state) means that the column analyzed has a nominal capacity P_{cap} equal to the AASHTO LRFD required design capacity or $P_{\text{cap}} = P_{\text{design}} = 15$ MN (3,404 kips). A value of 1.1 indicates that the capacity is 10% higher than the required design capacity such that $P_{\text{cap}} = 16.6$ MN (3,744 kips).

The steepness of the slope of each curve gives an assessment of the cost implied when an increase in the reliability index is desired. For example, a steep positive slope would indicate that large increases in the reliability index could be achieved as a result of small increases in the member capacities while a shallow positive curve would indicate that large increases in the member capacities would be required to achieve small increases in the reliability index.

62

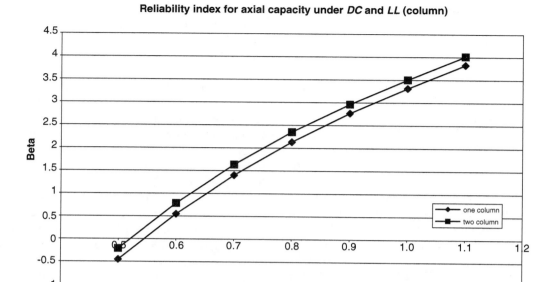

Figure 3.8. Reliability of concrete columns under axial load due to gravity.

The results in Figure 3.8 for the columns under axial load show that the current AASHTO LRFD produces reliability indexes of 3.50 for the two-column bent and 3.32 for the one-column bent. This indicates that the AASHTO LRFD code produces reliability index levels close to the 3.5 target beta set by the AASHTO LRFD code writers for columns under axial compression even though this particular limit state was not specifically considered during the calibration of the AASHTO LRFD specifications (Nowak, 1999). The figure also shows

that an increase in column capacity of 10% ($P_{cap}/P_{design} = 1.1$) would result in an average reliability index on the order of 3.90. This indicates that improvement in the reliability index can be achieved with relatively small increases in column axial capacities.

The results for the failure of the soil bearing capacity are illustrated in Figure 3.9. These results show that the current AASHTO LRFD produces reliability indexes of 2.53 for the two-column bent and 2.47 for the one-column bent, which

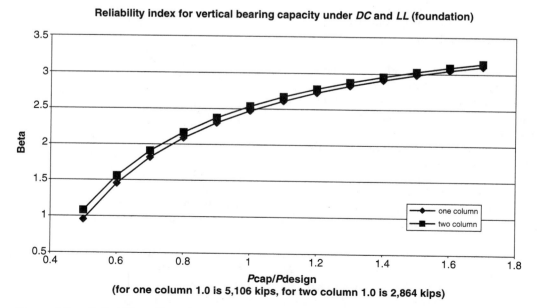

Figure 3.9. Reliability of foundation bearing capacity under axial load due to gravity.

indicates that the current code produces lower reliability levels for foundation design than for the design of structural elements despite the higher resistance factor of 0.50 used for the design of foundations. This is due to the high level of uncertainty associated with current methods for designing foundations and in estimating soil properties. The data are applicable to the models described by Poulos and Davis (1980). It should be noted that there are several different methods used in current practice for the design of foundations. These will produce different reliability indexes depending on the biases and implicit safety factors included when the methods are developed. NCHRP Project 12-55, entitled "Load and Resistance Factors for Earth Pressures on Bridge Substructures and Retaining Walls," is addressing some of these issues. Figure 3.9 shows that an increase in foundation bearing capacity of 50% ($P_{cap}/P_{design} = 1.5$) would result in an average reliability index on the order of 3.0 when P_{design} is calculated based on the Poulos and Davis method (1980). A much higher increase would be required to achieve a reliability index of 3.50. Hence, increases in the reliability indexes for bridge foundations would require large increases in the foundation depths and sizes because of the relative shallowness of the reliability index curve.

Figures 3.8 and 3.9 also show that the reliability indexes for the two- and one-column bents are reasonably similar despite the differences in bent configurations and column sizes.

Figure 3.10 illustrates how the reliability index against foundation overtipping for one-column bents varies with foundation depth. The figure shows that if the foundation is built to a depth of $L = 1.8$ m (6 ft) as required using current AASHTO LRFD design standards along with the Broms method for foundation analysis as described in Poulos and Davis (1980), the reliability index against overtipping is then equal to 3.58.

The reliability analysis is based on the free body diagram of Figure 3.3 and the random variables associated with soil resistance listed in Table 2.2.

It is noticed that doubling of the foundation depth from 1.8 m to 3.7 m (6 ft to 12 ft) would result in increasing the reliability index to 4.73. The higher reliability index for foundation overtipping compared with that for foundation bearing capacity is due to the fact that the Broms method has an implicit bias of about 1.50, as reported by Poulos and Davis (1980). On the other hand, the model proposed by Poulos and Davis (1980) for calculating the ultimate pile shaft bearing capacity for vertical load is not associated with any bias. The differences emphasize the need to develop more consistent reliability-based models for bridge foundation design. As mentioned earlier, NCHRP Project 12-55 is addressing some of these issues. The last failure mode considered for gravity loads only is the failure of the one-column bent because of bending. Although the live load eccentricity of the one-column bent is relatively small and the column will mostly behave as a member under axial loading, the possibility of bending failures is considered in this section producing the results presented in Figure 3.11. These results will be used in subsequent sections when studying the combination of live loads and other lateral loads that will produce high bending moments. If the column is designed to resist the applied moment produced because of the eccentricity of the live load, then the AASHTO LRFD specifications require a nominal moment capacity $M_{cap} = 8$ MN-m (5,908 kip-ft), as shown in Table 3.1. According to Figure 3.11, such a moment capacity would produce a reliability index of $\beta = 3.71$. This value is higher than the $\beta_{target} = 3.5$ used to calibrate the AASHTO LRFD and is also higher than the value observed in Figure 3.8 for column axial capacity. The difference is partially due to the fact that the dead load does not

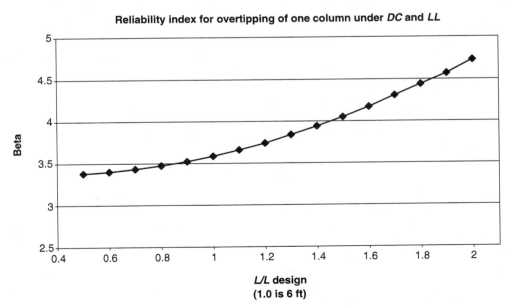

Reliability index for overtipping of one column under *DC* and *LL*

Figure 3.10.   Reliability for foundation overtipping caused by live and dead loads.

Reliability index for moment capacity for one column under live load

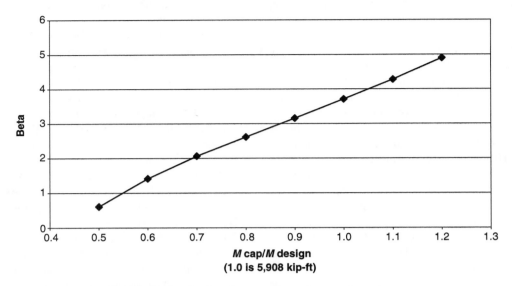

Figure 3.11. Reliability index for bending moment of one-column bent.

contribute in either resisting or increasing the bending moment in one-column bents. The slope of the reliability index curve shows that an increase in moment capacity by 10% would produce a reliability index equal to 4.28 (i.e., an increase in beta of more than 0.57).

The results presented in Figures 3.8 through 3.11 further illustrate that, based on the data assembled in Chapter 2, the AASHTO LRFD specifications have achieved a reasonably uniform reliability index close to the target value of 3.5 for structural members subjected to gravity loads. The reliability indexes for foundation design, however, greatly depend on the design methodology followed. Several different foundation design methodologies are used in practice that would require different safety factors to achieve a uniform reliability level. NCHRP Project 12-55 is currently addressing the issue of the foundation safety and the calibration of resistance factors for bridge foundation design. The slopes of the reliability index curves demonstrate that by increasing the structural properties of the bridge columns by 10%, a relatively large increase in the reliability index on the order of 0.40 to 0.60 is achieved. On the other hand, very large increases in the foundation capacities are required to effect small increases in the reliability index values associated with column overtipping or foundation bearing failures.

The results shown in Figures 3.8 through 3.11 show the reliability index values calculated for member failure. When the column in a single-column bent fails, it leads to the failure of the whole system. On the other hand, when one member of a multicolumn bent system fails in bending, the presence of ductility and redundancy will help the system redistribute the load to provide additional reserve strength. In this case, system failure will occur at a higher load. Failures in shear or com-

pression do not provide much ductility. Because under gravity load the failure of multicolumn bents is primarily due to axial compression and soil-bearing capacity rather than due to bending, all types of bents subjected to gravity loads will have no redundancy.

3.2.2 Reliability Analysis for Wind

The reliability analysis of the one-column and two-column bents under wind loads is also executed using the model described in Chapter 2 and using the free body diagrams of Figures 3.4 and 3.7 with the live load $F_{LL} = 0$. Referring to Figure 3.4(b), the failure function for column bending in the one-column bent can be represented by an equation of the following form:

$$Z_5 = M_{col} + \frac{3\gamma D K_p f^2}{2}\frac{f}{3}\lambda_{cyc} \\ - [F_{WL1}(e_1 + f) + F_{WL2}(e_2 + f)] \tag{3.18}$$

where

M_{col} = the column bending moment capacity,
F_{WL1} = the wind load on the superstructure,
F_{WL2} = the wind load on the column,
γ = the specific weight of the soil,
D = the column diameter,
K_p = the Rankine coefficient,
e_1 and e_2 = the distance of F_{WL1} and F_{WL2} from the soil level,
f = the distance from the soil level to the point of maximum moment, and

λ_{cyc} = the model of the effect of applying cyclic loads on the foundation.

The distance, f, is calculated by setting the sum of F_{WL1} + F_{WL2} = to the lateral load pressure, or

$$\frac{3\gamma D K_p f^2}{2} = F_{WL1} + F_{WL2} \qquad (3.19)$$

Referring to Equation 3.18, failure occurs when Z_5 is less than or equal to zero. All the variables used in Equations 3.18 and 3.19 are considered random except for the column diameter, D, and the eccentricities, e_1 and e_2. Adjustments to Equations 3.18 and 3.19 are made when analyzing the two-column bents to find the portion of F_{WL1} and F_{WL2} applied on each column. The statistical models used to describe the random variables are provided in Tables 2.1, 2.2, and 2.7. Specifically, the wind load F_{LW}, for either wind on the structure or on the column, is composed of time-dependent and time-independent random variables and can be represented as

$$F_{WL} = c C_p E_z G (\lambda_V V)^2 \qquad (3.20)$$

where

c = an analysis constant,
C_p = the pressure coefficient,
E_z = the exposure coefficient,
G = the gust factor,
λ_V = the statistical modeling factor for wind speeds, and
V = the wind speed.

All the terms in Equation 3.20 are random with statistical properties provided in Table 2.6 of Chapter 2. The wind speed, V, is a time-dependent random variable that varies as a function of the return period as described in Equations 2.11 and 2.14. Mean and COV values of wind speed at different interior sites throughout the United States are provided in Table 2.7 for the annual maximum winds and the 50-year maximum wind.

The reliability analysis is performed in this section to study the safety of bridge columns in bending caused by the effect of wind alone. The return period used is 75 years in order to remain compatible with the AASHTO LRFD criteria. Notice that Equation 3.18 does not include the effect of dead loads. Only the bending in the column failure mode is considered because preliminary calculations performed in this study have indicated that the safety of the one-column bent against overtipping is very high and thus overtipping of one-column bents because of wind loads is not considered in this report.

The results of the reliability index calculations for the failure of one column in bending are illustrated in Figure 3.12 for the one-column bent and Figure 3.13 for the two-column bents. The calculations are effected for different wind data collected by Simiu, Changery, and Filliben (1979) and summarized by Ellingwood et al. (1980) at seven sites within the interior of the United States. The results show that bridge structures designed to satisfy the current AASHTO LRFD specifications for wind loads (i.e., for M_{cap}/M_{design} = 1.0) produce an average reliability index value from all the sites analyzed equal to 3.07 for the one-column bent and 3.17 for the two-column bent. However, the variability of the results for

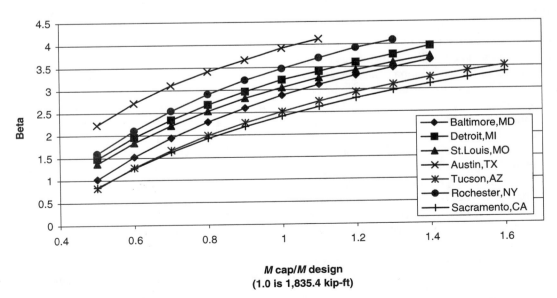

Reliability index for moment capacity of one-column bent

Figure 3.12. Reliability index for bending moment of one-column bent under wind loads.

Reliability index for moment capacity of column in two-column bent

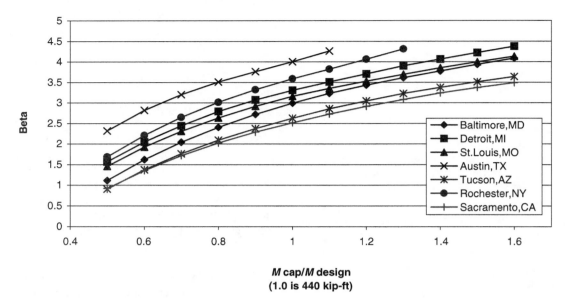

Figure 3.13.   Reliability index for bending moment in two-column bent under wind loads.

different sites is large: the range of the reliability index is from 2.41 to 3.92 for one-column bents and from 2.51 to 4.00 for two-column bents. This large spread in β is due to the fact that the ASCE-7 design wind speed maps that are adopted by AASHTO use a common design wind of $V = 145$ km/h (90 mph) for all interior regions; the actual data collected (see Table 2.7) show a large range of values for the mean maximum winds and their COVs. It is herein recommended that future wind maps account for the different wind speeds observed in different regions.

The results shown for the 7.6-m (25-ft) high bridge produce low applied moments from the wind load as compared with the moments caused by the effect of live load. Thus, a sensitivity analysis is performed to study the effect of changing the bridge height on the reliability of the bridge column. The results for failure of the one-column bent in bending are shown in Figure 3.14 for different column bents with heights varying between 7.6 m and 23 m (25 ft and 75 ft). The results illustrate that column height does not significantly affect the reliability index of bridges for the failure mode in bending although a small drop in the reliability index is observed as the column height increases. The drop in β between the heights of 7.6 m and 23 m (25 ft and 75 ft) is generally less than 0.30 such that the average reliability index from the three sites studied—namely St. Louis, Sacramento, and Austin—reduces from a value of 3.13 to an average value of 2.87. The M_{design} values used to normalize the abscissa of the curve shown in Figure 3.14 are the nominal bending moment capacities required for exactly satisfying the current AASHTO LRFD criteria. The drop in β is primarily due to the uncertainty in identifying the location of the point of maximum

moment, f, calculated from Equation 3.19. The value of f is highly uncertain because of the uncertainty in the soil properties and the modeling of the soil capacity to resist lateral forces. The drop in the reliability index caused by increases in column height is clearly less significant than the difference observed in the effects of site variations in wind speeds on the reliability index.

The results illustrated in Figures 3.12 through 3.14 show the reliability index values calculated for member failure. When the column in a single-column bent fails, it leads to the failure of the whole system. On the other hand, when one member of a multicolumn bent system fails in bending, the presence of ductility and redundancy will help the system redistribute the load to provide additional reserve strength. In this case, system failure would occur at a higher load. Liu et al. (2001) have shown that typical two-column bent systems formed by pile extensions subjected to lateral loads provide an additional reserve strength equal to 15% higher than the strength of one column. Using a system capacity ratio of 1.15 with a COV of 8% (as described in Table 2.1 for unconfined columns) would lead to higher system reliability index values as compared with those of one column. Figure 3.15 illustrates the difference between the reliability indexes for one column versus for the two-column bent system. The wind data used as the basis of the calculations are from the St. Louis site. The results show an increase from the 3.16 reliability index for one column to a value of 3.40 for the two-column system, or an increase of about 0.24. Higher increases in the reliability index would be expected if the concrete columns are highly confined because of the effect of the higher ductility levels associated with confined columns.

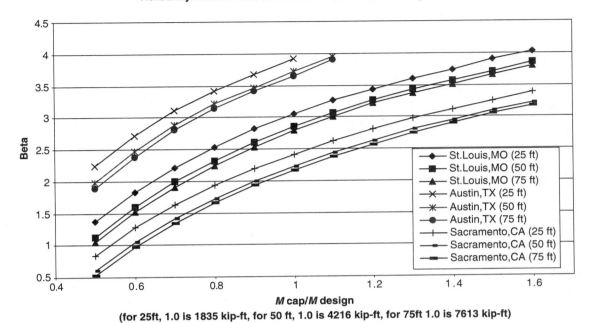

Figure 3.14. Reliability index for bending moment in one-column bent of different height under wind loads.

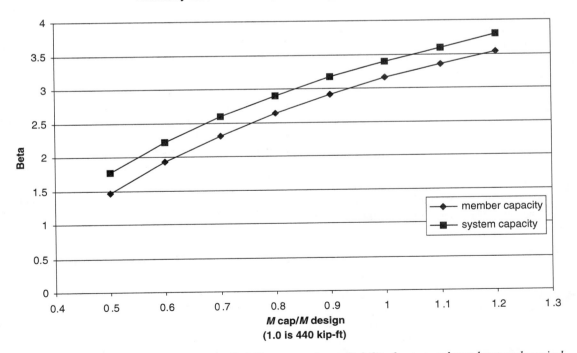

Figure 3.15. Comparison of system reliability to member reliability for two-column bent under wind load.

3.2.3 Reliability Analysis for Earthquake Alone

The reliability analysis of the one-column and two-column bents under earthquakes is also executed using the model described in Chapter 2 and using the free body diagrams of Figures 3.4 and 3.7 with the live load of $F_{LL} = 0$. For the case of earthquakes, two limit states are considered: (1) the bending of the columns of one- and two-column bents because of lateral inertial forces and (2) the overtipping of the one-column bent. Referring to Figure 3.4(a) and (b), the failure functions for column bending in the one-column bent can be represented by an equation of the following form:

$$Z_6 = M_{col} + \frac{3\gamma D K_p f^2}{2} \frac{f}{3} \lambda_{cyc} - F_{EQ}(e_1 + f) \quad (3.21)$$

where

M_{col} = the column bending moment capacity,
F_{EQ} = the equivalent earthquake lateral load on the column,
γ = the specific weight of the soil,
D = the column diameter,
K_p = the Rankine coefficient,
e_1 = the distance of F_{EQ} from the soil level,
f = the distance from the soil level to the point of maximum moment, and
λ_{cyc} = the model of the effect of cyclic loading on the foundation.

The distance f is calculated by setting the equivalent lateral earthquake load, F_{EQ}, equal to force, P_p, of Figure 3.4 produced from the earth pressure.

Based on Figure 3.4(a), the failure equation for overtipping can be represented as follows:

$$Z_7 = \frac{3\gamma D K_p L^2}{2} \frac{L}{3} \lambda_{cyc} - F_{EQ}(e_1 + L) \quad (3.22)$$

where L is the depth of the foundation. Notice that Equation 3.22 does not consider the counter effects of the dead weights. This is because it is herein assumed that vertical accelerations caused by the earthquake may significantly reduce the contributions of the dead weight in resisting the risk for overtipping.

Referring to Equations 3.21 and 3.22, failure occurs when either Z_6 or Z_7 are less than or equal to zero. All the variables used in Equations 3.21 and 3.22 are considered random except for the column diameter, D; the eccentricity, e_1; and foundation depth, L. The statistical models used to describe the random variables are provided in Tables 2.1, 2.2, and 2.6. Specifically, the equivalent lateral earthquake load, F_{EQ}, is composed of time-dependent and time-independent random variables and can be represented as follows:

$$F_{EQ} = \lambda_{EQ} C' S_a(t'T) \frac{AW}{R_m} \quad (3.23)$$

where

λ_{EQ} = an analysis modeling factor;
C' = the spectrum's modeling factor;
$S_a(\ )$ = the spectral acceleration calculated as a function of the calculated nominal period, T;
t' = the period's modeling factor;
A = the earthquake intensity in terms of the ground acceleration, g;
W = the weight of the structure; and
R_m = the response modification factor.

When studying overtipping, the response modification factor is $R_m = 1.0$, and it is deterministic. For column bending, R_m is random. The variables λ_{EQ}, C', and t' are also random. W is assumed to be deterministic although the uncertainties in estimating W are considered in λ_{EQ} as well as in t'. The earthquake intensity, A, is the only time-dependent random variable that varies as a function of the return period, as is described in Equations 2.11 and 2.14. The probability distribution of A is described in Figure 2.4 for five different sites. The probability of exceedance curves of Figure 2.4 are for the annual maximum earthquake intensity. The calculations performed in this report are executed for a 75-year design period in order to remain consistent with the AASHTO LRFD specifications.

The reliability analysis of one-column bent and two-column bent bridges has been performed using data from the five different sites listed in Table 2.4 and Figure 2.4. Figures 3.16 and 3.17 show the reliability index for the bending failure limit state for each of the five sites as a function of the column moment capacity for the one-column bent and the two-column bent, respectively. The abscissa of the plot is normalized such that a ratio of 1.0 indicates that the bridge is designed to exactly satisfy the proposed NCHRP Project 12-49 requirements [ATC and MCEER, 2002] with a nominal response modification factor $R_m = 1.5$. The design column moment capacities and foundation depths required to resist overtipping are provided in Tables 3.3 and 3.9 for five different site conditions. For the purposes of this study, the bridges are subjected to earthquake input motions similar to those expected in San Francisco, Seattle, St. Paul, New York, or Memphis. The reliability analysis is then performed for each bridge configuration and for each of the five site data assuming a 75-year design life. The object is to verify whether the proposed NCHRP Project 12-49 specifications would provide reasonably uniform reliability levels for any site within the United States.

Figure 3.16 shows that the proposed NCHRP Project 12-49 specifications using a nominal response modification factor of $R_m = 1.5$ will produce a reliability index, β, that varies between 2.56 and 2.88. The average from the five sites is equal to 2.78. If a response modification factor of $R_m = 1.0$ is used (such as proposed for columns with hinges that cannot be inspected), then the reliability indexes will have an average value of 3.00 with a range of 2.80 to 3.16. These values can be observed in Figure 3.16 for an M_{cap}/M_{design} ratio of 1.50

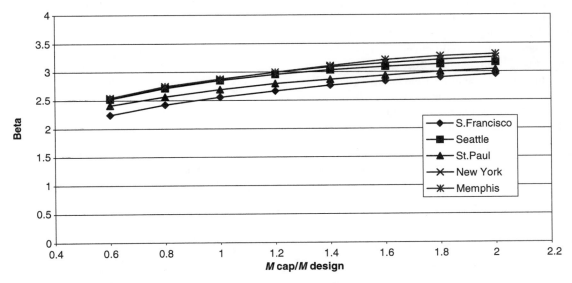

Figure 3.16. Reliability index for one-column bent under earthquakes.

because changing the nominal response modification factor will result in increasing the moment capacity by the same ratio. The small increase in the reliability index caused by the 50% change in M_{cap} reflects the high costs that would be required to produce only an incremental change in the reliability of bridges subjected to earthquakes. This is due to the fact that the uncertainties in estimating the earthquake intensities are very high.

Figure 3.17 shows that the range of the reliability indexes for the two-column bent is slightly higher than that of the one-column bent. The average reliability index for the five earthquake data is 2.82 with a minimum β equal to 2.70 and a maximum value equal to 2.91. Since member ductility is already taken into consideration by using the response modification factor, the reliability indexes shown in Figures 3.16 and 3.17 are for the complete system.

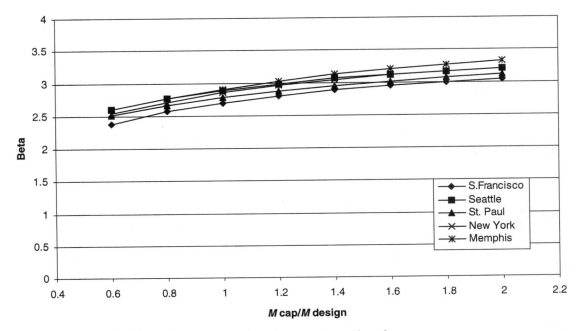

Figure 3.17. Reliability index for two-column bent under earthquakes.

The results shown in Figures 3.16 and 3.17 are for a foundation depth $L = 24$ m (80 ft). The effect of the foundation depth on the bridge safety is illustrated in Figures 3.18 and 3.19. Figure 3.18 shows how the reliability index for the bending limit state for the one-column bent varies with foundation depth. It is noticed that the effect is negligibly small. This is due to the fact that a shallow foundation would reduce the stiffness of the system, thus increasing its natural period, which in turn leads to lower spectral accelerations and lower moments in the column. The maximum difference between the reliability indexes for each site is less than 0.03. Similarly, it is observed that the depth of the foundation does not significantly affect the reliability index for the bending limit state of the two-column bent. The reliability indexes shown in Figure 3.19 range from a low value of 2.69 to a high of 2.95 for all foundation depths and site data.

Unlike what was observed for column bending, the depth of the foundation affects the safety of the column against overtipping, as is illustrated in Figure 3.20. In Figure 3.20, the reliability index is calculated using the failure function of

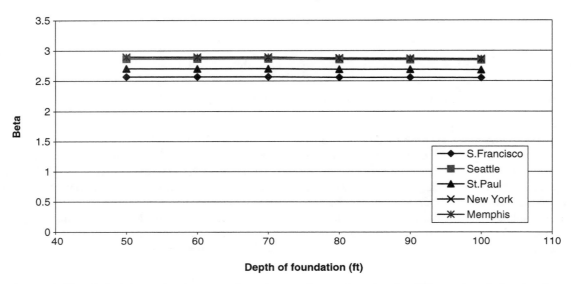

Figure 3.18.   Reliability index for one-column bent under earthquakes for different foundation depths.

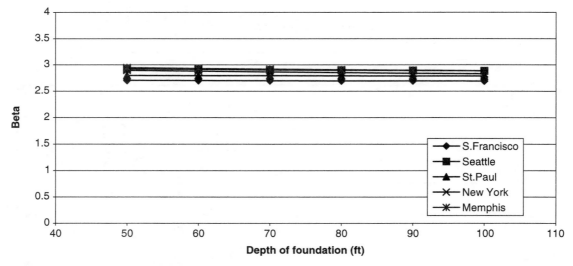

Figure 3.19.   Reliability index for two-column bent under earthquakes for different foundation depths.

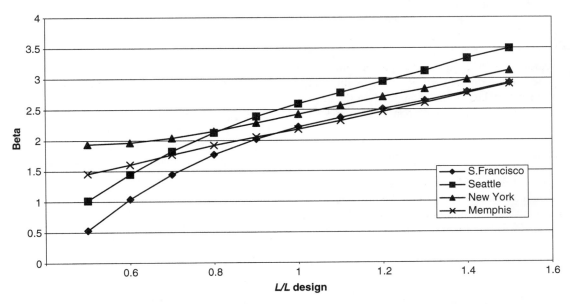

Figure 3.20. Reliability index for one-column bent for overtipping caused by earthquakes.

Equation 3.22 for different values of foundation depth, L. The required design foundation depth L_{design} is given in Table 3.3 for the different site data. The results show that the effect of the foundation depth becomes significant because the shallower foundations will not be able to resist the tendency of columns to overtip under the effect of the inertial forces at the top of the column. If the column depth is designed to satisfy the requirements of the proposed specifications of NCHRP Project 12-49 (ATC and MCEER, 2002), then the reliability index will vary between 2.22 and 2.59 for the sites analyzed in this report. It is noticed that the trends of the curves for the New York and Memphis data differ from those of San Francisco and Seattle because of the shallowness of the foundations that are required in the first two cases compared with the second set. The inertial forces observed for St. Paul are so small that the probability of overtipping is negligibly small. It is also noted that the results shown in Figures 3.16 through 3.19 ignore the possibility of vertical accelerations. These, according to the NCHRP Project 12-49, can be ignored except for near-field motions in which the vertical accelerations may be significant. The reliability calculations for overtipping did not include the vertical earthquake accelerations, nor did they account for the counter effect of the dead weight.

The analyses performed above are based on isolating a single bent from the bridge system that is founded on an extension pile and on modeling it as an SDOF system after determining its point of fixity. The reliability analysis is performed based on a design that uses a response modification of $R_m = 1.5$. Appendix H performs a reliability analysis of the columns of a bridge founded on a multiple-pile foundation system when the bridge is modeled as a multidegree-of-freedom (MDOF)

system. In this case, the bridge columns were designed using a design response modification factor $R_m = 6.0$ as recommended by NCHRP Project 12-49. The results of the analysis described in Appendix H demonstrate the following points:

1. The simplified analysis using an SDOF system as performed in this section yields values for the reliability index, β, that are consistent with those obtained from the more advanced multimodal structural analysis.
2. The use of a response modification factor $R_m = 6.0$ for bridge columns associated with the 2,500-year NEHRP spectrum produces system reliability index values on the order of 1.75, which is considerably lower than the 2.82 average value observed for multicolumn systems with members designed using $R_m = 1.5$. The 1.75 value that is for system safety is much lower than the member reliability equal to 3.5 that is used as the basis for calibrating the LRFD code for members under gravity loads.

In summary, large differences in the reliability indexes are observed for the different components of bridge subsystems subjected to seismic events. For example, the foundation systems produce much higher reliability index values than the bridge columns do. This difference has been intentionally built into the bridge seismic design process by seismic design code writers in order to account for the consequences of the failure of different members. However, the difference is rather large, producing a reliability index close to $\beta = 3.0$ when a response modification factor equal to $R_m = 1.0$ is used to about $\beta = 1.75$ when a response modification factor $R_m = 6.0$ is used during the design process. It is noted that the use

72

of a response modification $R_m = 1.0$ implies a design based on the elastic behavior of members while a value of $R_m = 6.0$ accounts for the plastic behavior of bridge columns. These differences, however, will not affect the results of the calibration of load factors for combinations of extreme events that include earthquakes as long as the target reliability level used during the calibration is based on the reliability index obtained when earthquakes alone are applied on the bridge. This point will be further elaborated upon in Section 3.3.1.

3.2.4 Reliability Analysis for Scour Alone

The scour model described in Chapter 2 can be used in a Monte Carlo simulation program to find the probability that the scour depth in a 75-year period will exceed a given value y_{CR}. In this case, the failure function is given as follows:

$$Z_8 = y_{CR} - y_{SC}. \tag{3.24}$$

The scour depth, y_{SC}, for rounded piers set in non-cohesive soils is calculated from

$$\ln y_{max} = -2.0757 + 0.6285 \ln D + 0.4822 \ln y_0 \\ + 0.6055 \ln V + \varepsilon \tag{3.25}$$

where

D = the pier diameter,
y_0 = the flow depth,
V = the flow velocity, and
ε = the residual error.

As explained in Chapter 2, D is a deterministic variable because the pier diameter can be accurately known even before the actual construction of the bridge; y_0 and V are random variables that depend on Manning's roughness coefficient, n, and the 75-year maximum discharge rate, which are random variables having the properties listed in Table 2.9. Based on the analysis of the residuals affected in Appendix I, ε may be considered to follow a normal distribution with mean equal to zero and a standard deviation equal to 0.406. The flow depth, y_0, and velocity, V, are calculated from the geometry of the channel (which is assumed to be deterministic), from the Manning roughness coefficient (n), and from the hydraulic discharge rate (Q) using the relationships provided in Section 2.4.4 of Chapter 2. In addition, the estimation of the discharge rate will be associated with statistical uncertainties represented by a statistical modeling random variable, λ_Q. The pertinent random variables and their properties are listed in Table 2.9 of Chapter 2. Only the discharge rate, Q, is a time-dependent random variable as it increases with longer return periods. Given the cumulative distribution for the maximum 1-year discharge rate, Q, the probability distribution can be calculated for different return periods using Equation 2.14. In the calculations performed in this report, a 75-year return period is used in order to remain compatible with the AASHTO LRFD. The calculations are executed for the discharge data from the five sites listed in Table 2.8.

The probability that y_{SC} will exceed a critical scour depth, y_{CR}, is calculated for different values of y_{CR} as shown in Figure 3.21 for the one-column pier for hydraulic discharge data collected from five different rivers. Figure 3.22 gives the same results for the two-column pier. The differences between

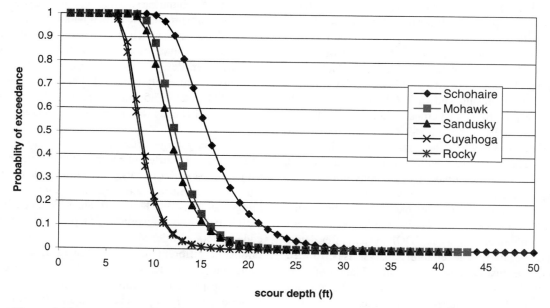

Figure 3.21.   Probability that actual scour will exceed critical depths for 6-ft-diameter column.

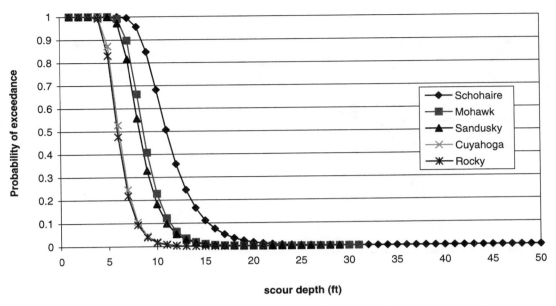

Figure 3.22. Probability that actual scour will exceed critical depths for 3.5-ft-diameter column.

the results of the two figures are due to the influence of the column diameter, *D*, in Equation 3.25.

Figures 3.23 and 3.24 show the reliability index, β, versus the normalized critical foundation depth such that a factor of 1.0 indicates that the foundation is designed to exactly satisfy the foundation depth requirement of HEC-18 for each of the river data considered. The safety of bridges against

scour is expressed in terms of the reliability index, β, that is obtained from

$$P_f = \Phi(-\beta) \qquad (3.26)$$

where, in this case, the probability of failure P_f is the probability that the scour depth, y_{SC}, will exceed the design scour

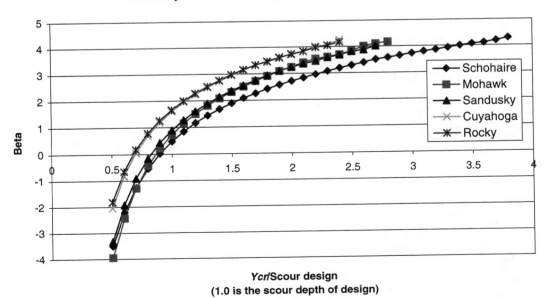

Figure 3.23. Reliability index for scour at different critical depths for D = 6 ft.

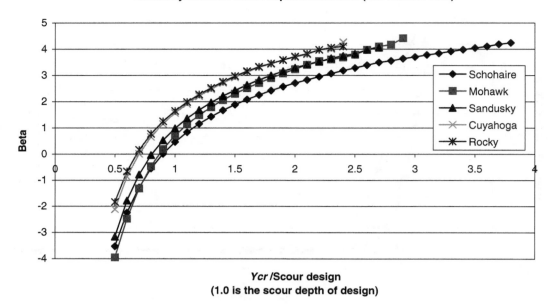

Figure 3.24. *Reliability index for scour at different critical depths for D = 3.5 ft.*

depth. Φ is the standard Gaussian cumulative distribution function.

The results shown in Figures 3.23 and 3.24 demonstrate that the HEC-18 method provides varying degrees of reliability for the different river data analyzed. The results of the simulation are also summarized as shown in Tables 3.12 and 3.13 for the five rivers. The two tables show the average 75-year discharge rate and the COV for the 75-year discharge rate for the one- and two-column bents. They also show the 75-year maximum scour depth along with its COV. These are com-

pared with the design scour depths as calculated from Equation 2.22 and listed in Tables 3.4 and 3.10.

The results in Tables 3.12 and 3.13 show that the reliability index, β, implied in current scour design procedures varies from about 0.47 to 1.66 with an average value of 1.08. These values are much lower than β = 3.5, which is the reliability index used as the basis for the calibration of the load factors for the combination of live and dead loads in the AASHTO LRFD specifications. Also, the target β = 3.5 is for member reliability while a foundation failure caused by scour will generally lead to the collapse of the complete system. The

TABLE 3.12 Summary of simulation results for one-column pier

River	Average Q_{75yr} (Q for 75 years) (ft³/sec)	COV of Q_{75yr}	Average y_{s75} (max. scour depth in 75 yrs)	COV of y_{s75}	Design depth (ft)	Reliability index, β
Schohaire	85,000	29%	16.3	25%	17.3	0.48
Mohawk	34,000	12%	12.6	21%	14.0	0.73
Sandusky	38,000	18%	12.0	22%	14.3	0.90
Cuyahoga	20,000	16%	8.88	21%	12.3	1.60
Rocky	21,000	19%	8.68	22%	12.3	1.66

TABLE 3.13 Summary of simulation results for two-column pier

River	Average Q_{75yr} (Q for 75 years) (ft³/sec)	COV of Q_{75yr}	Average y_{s75} (max. scour depth in 75 yrs)	COV of y_{s75}	Design depth (ft)	Reliability index, β
Schohaire	85,000	29%	11.6	25%	12.3	0.47
Mohawk	34,000	12%	8.95	21%	9.93	0.71
Sandusky	38,000	18%	8.57	22%	10.2	0.99
Cuyahoga	20,000	16%	6.33	21%	8.70	1.57
Rocky	21,000	19%	6.18	22%	8.75	1.65

results show that the rivers with lower discharge rates have higher reliability levels than do those with the higher discharge rates. This confirms the observation made in Appendix I that the "design" scour depths as calculated from the HEC-18 equations give different levels of safety (or bias) when compared with observed scour depths with the safety level decreasing as the observed scour depth increases. In addition, the low overall reliability level observed for foundations designed to exactly satisfy the HEC-18 equations imply that although the scour design equation involves a safety factor equal to 2.0 (the first term in Equation 2.30), the large differences between the observed scour depths and those predicted (see Figure 2.10) lead to an overall low reliability index for scour as compared with the reliability index for bridge structural members. Based on this observation, it is recommended that a revision of current scour design procedures be initiated in order to improve our ability to predict the depth of scour and to increase the safety levels of bridges that may be subject to foundation erosion caused by scour. An increase in the safety levels for scour design is also justified based on the observation by Shirole and Holt (1991), who have reported that a majority of the bridges that have failed in the United States and elsewhere have failed because of scour. In fact, Shirole and Holt state that over a 30 year-period, more than 1,000 of the 600,000 U.S. bridges have failed and that 60% of these failures are due to scour. This would constitute a failure rate of 0.001 in 30 years, or 0.0025 failures in 75 years. However, since many of the U.S. bridges are not founded in water channels, the data reported by Shirole and Holt indicate a very high probability of failure for bridges founded in water channels. If one guesses that 12,500 bridges are exposed to scour, then the rate of failure would be about 12%. A reliability index of about 1.10 would mean a probability of failure of roughly 13.6% in 75 years.

Tables 3.14 and 3.15 give the additional load factor that should be applied on the scour design equation in order to produce different values of the reliability index when Equation 2.22 is used to determine the design scour depth. According to this approach, the results for y_{max} obtained from Equation 2.22 should be further multiplied by an additional load factor to produce reliability index values compatible with those observed with the other extreme events studied in this

report. For example, if a target reliability index of $\beta_{target} = 3.50$ is desired, then the design scour depth that should be used when designing bridge foundations for scour should be equal to 2.15 times the value calculated from Equation 2.22. It is noted that two of the rivers selected have relatively low discharge volumes, two rivers have average discharge volumes and only one river has a relatively high discharge volume. If a weighted average is used, then a factor of 2.24 should be used rather than the 2.15 calculated above.

On the other hand, if a reliability index of $\beta_{target} = 2.50$ is to be used as the target index for the design of bridge foundations for scour, then the average load factor should be equal to 1.52 while the weighted average would be 1.57. For a target of $\beta_{target} = 3.00$, the average load factor is 1.79 while the weighted average would be 1.87. It is important to note that significant differences are observed in the load factors calculated from different sites. On the other hand, no difference is observed for different column diameters. Sensitivity analyses performed in Appendixes C and D have also shown that variations in the shape and geometry of the water channel do not seem to influence the reliability index value significantly. Some of the other extreme events have shown lower average reliability indexes than 3.5. For example, wind loads have an average reliability index value on the order of 3.12 for the 25-ft-high columns, and the reliability index for the failure of columns in bending to ship collisions is found to be on the order of 2.78. Hence, for the sake of keeping the safety levels relatively uniform for the different extreme events considered, it would be justified to use a target reliability index of 3.0 for scour. It is noted that the reliability calculations for seismic events produced a reliability index even lower than those of the other events. However, as mentioned earlier, the costs associated with increasing the safety level for earthquake effects have prevented the use of the higher safety levels for earthquakes. Because the AASHTO LRFD specifications are primarily directed toward short- to medium-span bridges, which are normally set over small rivers with relatively low discharge rates, it is herein concluded that the use of a scour factor on the order of $\gamma_{sc} = 2.00$ would provide reliability levels slightly higher than the 3.0 target that is similar to the average reliability index observed for the other extreme events. The 2.00 value proposed herein is a rounded up value to the 1.87 value calculated above.

TABLE 3.14 Design scour depths required for satisfying different target reliability levels for one-column pier

Target index	β = 3.5		β = 3.0		β = 2.5	
Required depth (ft)	Depth	Load factor	Depth	Load factor	Depth	Load factor
River						
Schohaire	46.5	2.69	38.6	2.23	32.0	1.85
Mohawk	30.2	2.16	25.9	1.85	22.3	1.59
Sandusky	31.9	2.23	26.3	1.84	22.5	1.57
Cuyahoga	22.4	1.82	18.8	1.53	16.0	1.30
Rocky	22.5	1.83	18.6	1.51	15.9	1.29
Average		*2.15*		*1.79*		*1.52*

TABLE 3.15 Design scour depths required for satisfying different target reliability levels for two-column pier

Target index	β = 3.5		β = 3.0		β = 2.5	
Required depth (ft)	Depth	Load factor	Depth	Load factor	Depth	Load factor
River						
Schohaire	33.5	2.72	27.6	2.24	22.8	1.85
Mohawk	21.4	2.16	18.5	1.86	15.9	1.60
Sandusky	22.2	2.18	18.4	1.80	15.6	1.53
Cuyahoga	16.0	1.84	13.4	1.54	11.4	1.31
Rocky	16.0	1.83	13.2	1.51	11.3	1.29
Average		*2.15*		*1.79*		*1.52*

3.2.5 Reliability Analysis For Vessel Collision Forces

The reliability analysis of the one-column and two-column bents is also executed for vessel collision forces following the model described in Section 2.4.5 of Chapter 2. Using the free body diagram of Figure 3.4(b) and a force of $F_2 = F_{CV}$ where F_{CV} is the vessel collision force, two limit states are considered: (1) the bending of the columns of one- and two-column bents because of lateral collision force and (2) the shear failure of the impacted column. Referring to Figure 3.4(b), the failure function for column bending in the one-column bent can be represented by an equation of the form

$$Z_9 = M_{col} + \frac{3\gamma D K_p f^2}{2} \frac{f}{3} \lambda_{cyc} - F_{CV}(e_2 + f) \qquad (3.27)$$

where

M_{col} = the column bending moment capacity,
F_{CV} = the vessel collision force on the column,
γ = the specific weight of the soil,
D = the column diameter,
K_p = the Rankine coefficient,
e_2 = the distance of F_{CV} from the soil level,
f = the distance from the soil level to the point of maximum moment, and
λ_{cyc} = the model of the effect of cyclic loading on the foundation.

The distance f is calculated by setting the equivalent vessel collision force, F_{CV}, equal to the force from the soil pressure on the foundation, P_p, of Figure 3.4 produced from the earth pressure. For shearing failure of the column, the failure function is as follows:

$$Z_{10} = V_{col} - F_{CV} \qquad (3.28)$$

where V_{col} is the shearing capacity. The vessel collision force is calculated from

$$F_{CV} = x\,w\,P_B \qquad (3.29)$$

where x is the vessel collision modeling factor, and w is a factor that accounts for the statistical modeling of the calculated impact force, P_B. All the variables listed in Equations 3.27 through 3.29 are random except for the column diameter, D, and the distance, e_2. The statistical properties for all the random variables are listed in Tables 2.1, 2.2, and 2.12. P_B is the only time-dependent variable. The cumulative probability distribution of the maximum value of P_B in 1 year is provided in Figure 2.13 of Chapter 2. The calculations executed in this report are for a 75-year return period. The probability distribution for a 75-year return period for P_B is found by using Equation 2.14.

The results of the reliability calculations are shown in Figures 3.25 and 3.26. As shown in Figure 3.25, the reliability index for shearing failure of the column is calculated to be 3.15 when the column is designed to exactly meet the requirement of the AASHTO LRFD specifications for vessel collisions. A 10% increase in the column shearing capacity will increase the reliability index by 0.35, leading to a reliability index of 3.5. This demonstrates that it would not be very

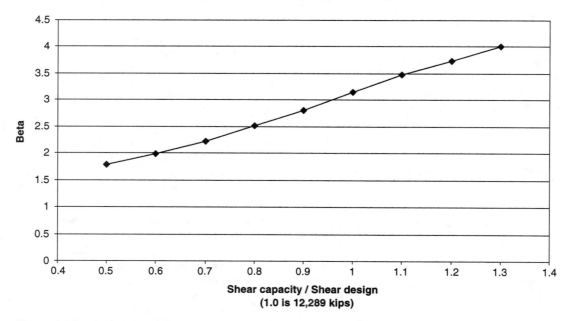

Reliability index for shear capacity under flotilla collisions

Figure 3.25.   Reliability index for shearing failure caused by vessel collisions.

Reliability index for moment capacity under ship collision

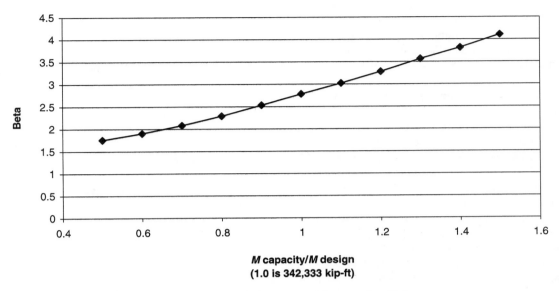

Figure 3.26. Reliability index for bending failure caused by vessel collisions.

costly to increase the target reliability level for shearing failures caused by vessel collisions. For column bending (see Figure 3.26), the reliability index is found to be equal to 2.78 if the column is exactly designed to satisfy the AASHTO LRFD specifications. A 10% increase in the column moment capacity will lead to a 0.24 improvement in β. The lower reliability index for moment failures is due to the inherent biases and conservatism implied in the shear design equations.

The reliability calculations executed in this section for each of the extreme events will be used in Section 3.3 to determine the target reliability levels and, subsequently, the load factors for the combination of extreme events.

3.3 RELIABILITY ANALYSIS FOR COMBINATIONS OF EXTREME EVENTS AND CALIBRATION OF LOAD FACTORS

This section presents the results of the reliability analysis executed for combinations of extreme events. The results are also used to perform the calibration of the load factors for combinations of events. Specifically, the section studies the following combinations:

1. Combinations involving live loads:
   – Earthquakes, *EQ*, and live loads, *LL*;
   – Wind loads, *WS*, plus live loads, *LL*;
   – Scour, *SC,* and live loads, *LL*.
2. Combinations involving scour:
   – Earthquakes, *EQ*, and scour, *SC*;
   – Wind loads, *WS*, plus scour, *SC*;

   – Collision of vessels and scour, *SC,* in addition to the case of live load plus scour addressed earlier.
3. Combinations involving vessel collision forces:
   – Wind loads, *WS*, plus vessel collision forces, *CV*;
   – Wind loads, *WS*, plus collision of vessels, *CV*, and scour, *SC*.

Because the results obtained in Section 3.2 have shown very little difference between the reliability indexes of two-column bents and one-column bents when these columns are designed to satisfy the same AASHTO LRFD specifications, this section will concentrate on studying the reliability of one-column bents.

3.3.1 Combination of Earthquakes and Live Loads (*EQ + LL*)

The single-column bent described in Figure 3.1 is analyzed to illustrate how the combined effects of earthquake and live loads will affect the bent's reliability. The data from the five earthquake sites with probability distribution curves described in Figure 2.4 are used. The live load data are obtained from the models developed by Nowak (1999) under NCHRP Project 12-33 as described in Section 2.4.1 (Table 2.3). The reliability calculations follow the Ferry-Borges model described in Section 2.5. The following conservative assumptions are made:

- All earthquakes last 30 sec (½ min), during which time the moment at the base of the column remains at its highest value. The intensity of the earthquake response is constant as shown in Figure 2.15 for the ½-min duration of the earthquake.

- The number of earthquakes expected in 1 year are 8 for the San Francisco site data, 2 for Seattle, 0.50 for Memphis, 0.40 for New York, and 0.01 for St. Paul. This means that the expected number of earthquakes in a 75-year return period in San Francisco will be 600, 150 earthquakes for Seattle, 38 in Memphis, 30 in New York, and 1 in St. Paul.
- The probability distribution of the maximum yearly earthquake may be used to find the probability distribution for a single event using Equation 2.14.
- The live load model for the applied vertical load has the same biases as those provided in Table B-16 of Nowak's report (1999) for the mean maximum negative moments of two equal continuous spans. Particularly, the results provided by Nowak for the 80-ft spans are used as the basis for the calculations. These results show that for each one-lane loading event, the load effect is on the average 0.79 times the load effect obtained from the HL-93 loading configuration of the AASHTO LRFD specifications for one lane of traffic with a COV of 10%. For two lanes, the average load effect is 1.58 times the effect of one lane of HL-93 with a COV of 7%. These results assume 1,000 single-lane heavy truck events in 1 day and 67 two-lane truck events in 1 day based on the assumptions of Nowak (1999) and Moses (2001).
- Equation 2.11 is used to find the probability distribution of the live load for different return periods. In particular, the live load for a $t = \frac{1}{2}$ min period is calculated for combination with the earthquake load effects when an earthquake is on.
- The effects of each earthquake are combined with the effects of the $\frac{1}{2}$ min live load magnitude and projected to provide the maximum expected combined load in the 75-year bridge design life period using the Ferry-Borges model described in Section 2.5. The calculations account for the cases in which the earthquake occurs with the $\frac{1}{2}$ min live load and the cases in which the live load arrives when no earthquake is on (which constitute most of the time).
- The reliability calculations are executed for a column bending failure limit state.
- The reliability calculations account for the uncertainties associated with predicting the EQ intensity, estimating the bridge response given an EQ intensity, the uncertainty in projecting the live load magnitude, and the uncertainty in estimating the column capacities.
- The failure equation for bending of the column under the combined effect may be represented as follows:

$$Z = M_{\text{col}} + \frac{3\gamma D K_p f^3}{6} - M_{EQ+LL,75} \qquad (3.30)$$

where

M_{col} = the column's moment capacity;
γ = the specific weight of the soil;

D = the column diameter;
K_p = the Rankine coefficient;
f = the distance below the soil level at which point the maximum bending moment occurs; and
$M_{EQ+LL,75}$ = the applied moment caused by the combined effects of the live load, LL, and the earthquake load, EQ, in a return period of $T = 75$ years.

The maximum effect of the combined load can be represented as follows:

$$M_{EQ+LL,75} = \max \left\{ \begin{array}{c} \max_{n_1}[c_{EQ}A + \max_{n_2}[c_{LL}I_{LL}]] \\ \max_{n_3}[c_{LL}I_{LL}] \end{array} \right\} \qquad (3.31)$$

where

n_1 = the number of earthquakes expected in a 75-year period;
n_2 = the number of live loads expected in a $\frac{1}{2}$ min period;
n_3 = the number of live load events expected in a period equal to 75 years minus the times when an earthquake is on (i.e., n_3 = number of live load events in 75 years $- n_1 \times \frac{1}{2}$ min, which is almost equal to the number of truck events in 75 years);
A = the earthquake intensity for one event;
I_{LL} = the intensity of the live load for one event; and
c_{EQ} and c_{LL} = the analysis coefficients that convert the earthquake intensity and the live load intensity to moment effects (e.g., Equations 3.16 and 3.17 show how to convert the intensity of the live load into a moment effect, and Equations 3.22 and 3.23 show how to convert the intensity of the earthquake acceleration into a moment effect).

The results of the reliability analysis are provided in Figure 3.27 for the San Francisco earthquake data for different values of the column capacity to resist applied bending moment. The results obtained by considering the reliability of the bridge structure when only earthquakes are considered—that is, by totally ignoring the effects of live loads—are also illustrated in Figure 3.27. The plot illustrates how, for this site, the effects of earthquakes dominate the reliability of the bridge when subjected to the combined effects of live loads and earthquakes. The results for the San Francisco site, as well as results for the other sites, are summarized in Table 3.16. As an example, if the bridge column is designed to satisfy the proposed revised AASHTO LRFD specifications (the seismic provisions)

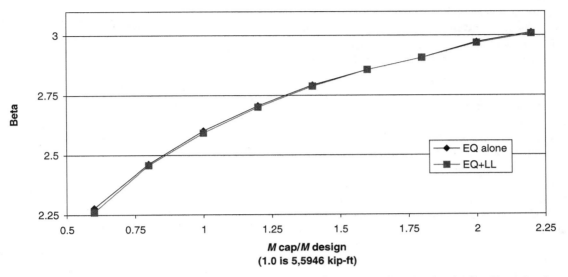

Figure 3.27. Reliability index for moment capacity under earthquakes plus live loads (San Francisco).

developed under NCHRP Project 12-49 (ATC and MCEER, 2002), then the required moment capacity for the drilled column shaft using a 2,500-year earthquake return period and a response modification factor equal to $R_m = 1.5$ will be $M_{cap} = M_{design} = 76$ MN-m (55,950 kip-ft). This will produce a reliability index (for a 75-year bridge design life) equal to $\beta = 2.60$ if no live load is considered. If one considers that a live load may occur within the ½ min when an earthquake is actively vibrating the bridge column, then the reliability index for the 76 MN-m column reduces to $\beta = 2.59$. The reduction is small. However, if one wishes to increase the reliability index from 2.59 back to the original 2.60, then the moment capacity of the column should be increased. By interpolation, the moment capacity that will produce a reliability index equal to $\beta = 2.60$ when both earthquake loads and live loads are considered should be $M_{cap} = 77$ MN-m (56,840 kip-ft) or an increase of 1207 kN-m (890 = 56,840 − 55,950 kip-ft).

The M_{design} value 76 MN-m (55,950 kip-ft) includes the effect of the resistance factor for bending, $\phi = 0.90$. The moment calculated from the applied earthquake force is equal to $M_{EQ} = 68$ MN-m (50,355 kip-ft). Also, the applied live load from the HL-93 live loading will produce a moment in

the column equal to $M_{LL} = 4,100$ kN-m (3,048 kip-ft). The live load factor that should be used for the combination of earthquakes and live load should be determined such that

$$\phi M_{req} = \gamma_{EQ} M_{EQ} + \gamma_{LL} M_{LL} \qquad (3.32)$$

or

$$0.90 M_{req} = \gamma_{EQ}\, 68 \text{ MN-m} + \gamma_{LL}\, 4100 \text{ kN-m.}$$

The calibration of the load combination factor involves the determination of the values of γ_{EQ} and γ_{LL} needed to produce the required column moment capacity $M_{req} = 77$ MN-m (56,840 kip-ft). If the value for γ_{EQ} is preset at 1.0, then γ_{LL} is calculated to be equal to 0.26. Many other options are available to reach the required 77 MN-m (56,840 kip-ft). However, in this example, it is assumed that the load factor for the moment obtained from the analysis of the earthquake effects is always equal to $\gamma_{EQ} = 1.0$. It is noted that in the examples solved in this section, the dynamic properties of the bridge system—particularly, the mass of the system—are not altered because of the presence of live load. The presence of live

TABLE 3.16 Summary of live load factors for combination of *EQ* plus *LL*.

Site	M_{req} (kip-ft)	M_{EQ}	M_{LL}	γ_{LL} for *LL+EQ*
San Francisco	56,840	50,351	3,048	0.26
Seattle	28,855	25,488	3,048	0.16
Memphis	21,597	18,945	3,048	0.17
New York	7,285	5,830	3,048	0.24
St. Paul	4,826	1,503	3,048	0.93
Average				*0.21*

loads on the bridge will increase the mass applied on the structure. However, since the vehicles may slide because of the effect of the earthquake, not all the mass will actually be active. Hence, the results shown in Table 3.16 assume that the mass of the system remains essentially constant despite the presence of the live load. A sensitivity analysis is performed further below to check the effect of this assumption on the final results.

Similar calculations are executed for all the five sites analyzed in this report—namely, San Francisco, Seattle, Memphis, New York, and St. Paul. The results from the five sites are summarized in Table 3.16. Notice that for all the sites considered except for St. Paul, the live load factor remains below 0.26 with an average value of 0.21. Based on the results shown in Table 3.16, it would seem appropriate to recommend that a live load factor of $\gamma_{LL} = 0.25$ associated with an earthquake factor of $\gamma_{EQ} = 1.00$ be used to account for the combination of live loads and earthquake loads for typical bridge bents supporting medium-span bridges. These proposed load factor values are on the lower side of the range of $\gamma_L = 0.25$ to 0.50 proposed by Nowak (1999) for heavy traffic sites (annual daily truck traffic = 5,000 trucks per day or 1,000 heavy trucks per day). Nowak suggests that lower values should be used for sites with low traffic volume and longer span lengths. (It is noted that only a range of values and no specific values are provided in the preliminary work of Nowak [1999].) Longer spans will produce lower live load factors, because as the span length increases, the mass of the bridge becomes dominant compared with the applied live loads and the contributions of the live loads become less significant. Assuming that all other parameters remain constant, a higher mass will produce higher dynamic forces.

The results for the St. Paul site, which produces a load factor $\gamma_{LL} = 0.93$, are removed from consideration because for this site, the live load dominates the design. In fact, if the bridge is designed for live loads alone, the required moment capacity is 8 MN-m (5,927 kip-ft), which is higher than the moment capacity required to produce a reliability index of 2.88 for the combined effects of earthquakes and live loads.

The analysis performed above ignored the effect of the truck's mass on the dynamic response of the bridge system. This assumption is justified because the truck is not rigidly attached to the structure. However, the existence of friction between the truck tires and the bridge deck may require that at least a portion of a truck's mass would contribute to changing the dynamic properties of the system. Because, to the knowledge of the authors, no information is available in the literature about the mass contributions of trucks to the dynamic response of bridges, a value of 20% of the truck masses is included in order to study how the load factors might change. This 20% value has been selected based on a rule of thumb that some seismic engineers have used in the past. The reliability analysis and the load calibration process is then repeated following the same model described above to yield the results described in Table 3.17. The results shown in Table 3.17

TABLE 3.17 Summary of live load factors considering contributions from truck masses to dynamic properties of system

Site	M_{req} (kip-ft)	M_{EQ}	M_{LL}	γ_{LL} for LL+ EQ
San Francisco	57,137	50,590	3,048	0.27
Seattle	29,274	25,616	3,048	0.24
Memphis	21,835	19,015	3,048	0.21
New York	7,347	5,856	3,048	0.25
~~St. Paul~~	~~4,826~~	~~1,529~~	~~3,048~~	~~0.92~~
Average				*0.24*

demonstrate that although the required load factor is slightly higher than that factor observed in Table 3.16, it remains close to the $\gamma_{LL} = 0.25$ value recommended above.

The calibration executed above is for the foundation system that was designed using a response modification factor $R_m = 1.5$. To study the effects of the different response modification factors that may be used for different bridge components, the calibration process is repeated assuming that a response modification factor $R_m = 6.0$ is used for the design of the bridge column. In this case, as explained in Appendix H, the reliability index, β, for the bridge subjected to earthquakes alone is lower than that obtained when $R_m = 1.5$ by more than 1.0. Using the lower reliability index as the target that should be reached when combining earthquake and live load effects would yield the results shown in Table 3.18. The results reported in Table 3.18 clearly show that the live load factor that would be required to maintain the same reliability index as that obtained when $R_m = 6$ is used with earthquakes alone remains essentially similar to that shown in Table 3.17. In Table 3.18, the results for the New York City site, however, must also be removed because when using $R_m = 6$, the required moment capacity for the bridge column will be dominated by the effects of the live load alone, as explained above for the St. Paul site.

Another sensitivity analysis is performed to study how the combination of load factors would change if the load factor for earthquakes is not preset at $\gamma_{EQ} = 1.0$. This is achieved by using an optimization algorithm with an objective function set to minimize the sum of the squared difference between

TABLE 3.18 Summary of live load factors considering contributions from truck masses to dynamic properties of system for bridge columns designed using $R_m = 6.0$

Site	M_{req} (kip-ft)	M_{EQ}	M_{LL}	γ_{LL} for LL + EQ
San Francisco	12,189	10,275	3,048	0.23
Seattle	6,889	5,412	3,048	0.26
Memphis	5,296	4,085	3,048	0.22
~~New York~~	~~4,114~~	~~1,339~~	~~3,048~~	~~0.78~~
~~St. Paul~~	~~3,931~~	~~362~~	~~3,048~~	~~1.04~~
Average				*0.24*

the results of Equation 3.32 and those of M_{req} that are given in the first column of Table 3.16 when M_{EQ} and M_{LL} are plugged into Equation 3.32 with unknown values of γ_{EQ} and γ_{LL}. The results show that the square of the difference would be minimized when the combination $\gamma_{EQ} = 1.00$ and $\gamma_{LL} = 0.18$ is used. This is only slightly different than the results shown in Table 3.16. Hence, for the sake of conservatism, the combination $\gamma_{EQ} = 1.00$ with $\gamma_{LL} = 0.25$ is recommended for use when designing bridges that are susceptible to threats from the combined effects of earthquakes and live loads. Please note that these results are very conservative given the previous assumptions that the earthquake is assumed to last for 30 sec at its peak value and the live load model follows the conservative assumptions described by Nowak (1999).

3.3.2 Combination of Wind and Live Loads (WS + LL)

The single-column bent described in Figure 3.1 is analyzed to illustrate how the combined effects of wind and live loads will affect the bent's reliability. The wind data from three sites—St. Louis, Austin, and Sacramento—are selected as representative for low, medium, and high wind intensity sites. Table 2.7 of Chapter 2 provides the basic wind intensity data for the three sites. The live load data are obtained from the models developed by Nowak (1999) for NCHRP Project 12-33 as described in Section 2.4.1 of Chapter 2 (Table 2.3). The reliability calculations follow the Ferry-Borges model described in Section 2.5. The following assumptions are made:

- All winds last for 4 h, during which time the moment at the base of the column remains at its highest value. The intensity of the wind response is constant, as is shown in Figure 2.15 for the 4-h duration of the wind.
- The winds are independent from each other.
- Each site is on the average exposed to 200 winds in 1 year.
- The probability distribution of the maximum wind speed follows a Gumbel distribution, which may be used to find the probability distribution for a single event using Equation 2.14.
- The live load model assumes that for each one-lane loading event, the load effect is on the average 0.79 times the load effect obtained from the HL-93 loading configuration for one lane of traffic with a COV of 10%. For two lanes, the average load effect for one event is 1.58 times the effect of one lane of HL-93 with a COV of 7%.
- Each site is exposed to 1,000 single-lane heavy truck events and 67 two-lane truck events in 1 day based on the assumptions of Nowak (1999) and Moses (2001).
- Equation 2.11 is used to find the probability distribution of the live load for different return periods. In particular, the live load for a 4-h period is calculated for combination with the wind load effects when a windstorm is on.

- The effect of each wind is combined with the effect of the 4-h live load magnitude and is projected to provide the maximum expected combined load in the 75-year bridge design life period using the Ferry-Borges model described in Section 2.5. The calculations account for the cases in which the wind occurs with the 4-h live load and the cases in which the live load arrives when no wind is on.
- The reliability calculations are executed for a column bending failure and column overtipping limit states.
- The reliability calculations account for the uncertainties associated with predicting the wind speed intensity, determining the bridge response given a wind load, the uncertainty in projecting the live load magnitude, and the uncertainty in estimating the column capacities and soil resistance properties.
- The failure equation for bending may be represented as follows:

$$Z = M_{col} + \frac{3\gamma D K_p f^3}{6} - M_{WS+LL,75} \qquad (3.33)$$

where

M_{col} = the column's moment capacity;
γ = the specific weight of the soil;
D = the column diameter;
K_p = the Rankine coefficient,
f = the distance below the soil level at which the maximum bending moment occurs; and
$M_{WS+LL,75}$ = the applied moment caused by the combined effects of the live load, LL, and the wind load, WS, in a return period of $T = 75$ years.

The maximum effect of the combined load can be represented as follows:

$$M_{WS+LL,75} = \max \left\{ \begin{matrix} \max\limits_{n_1}\left[c_{WS}V^2 + \max\limits_{n_2}[c_{LL}I_{LL}] \right] \\ \max\limits_{n_3}[c_{LL}I_{LL}] \end{matrix} \right\} \qquad (3.34)$$

where

n_1 = the number of winds expected in a 75-year period ($n_1 = 200 \times 75$);
n_2 = the number of live loads expected in a 4-h period when the wind is on ($n_2 = 167$ for one-lane loading and $n_2 = 11$ for two-lane loading);
n_3 = the number of live load events expected in a period equal to 75 years minus the times when a windstorm is on (i.e., $n_3 =$ number of live load events in 75 years − $n_1 \times 4$ h);

V = the wind speed for one windstorm;

I_{LL} = the intensity of the live load for one event; and

c_{WS} and c_{LL} = the analysis coefficients that convert the wind speed and the live load intensity to moment effects (e.g., Equations 3.16 and 3.17 show how to convert the intensity of the live load into a moment effect, and Equations 3.18 through 3.20 show how to convert the wind speed into a moment effect).

The effect of wind load on moving trucks is taken into consideration during the simulation. However, following the suggestion of the AASHTO LRFD specifications, the effect of the wind loads on live loads, and the effects of live loads in combination with wind loads are considered only if the wind speed at 10 m (30.5 ft) is less than 90 km/h (56 mph). For the cases in which the wind speed exceeds 90 km/h (56 mph), only the wind load on the structure is considered. The justification is that there is no truck traffic under extreme windstorms.

The results of the reliability analysis are provided in Figure 3.28 for the bending limit state of the 7.6-m (25-ft) bridge bent. Figure 3.29 gives the results for overtipping. The data are presented for the three representative wind sites for combination of wind loads and live loads and for wind loads alone. The effect of the soil in resisting bending and overtipping are included in the reliability analysis. Similarly, the contributions of the permanent load in resisting overtipping

are also included. The same calculations are performed for a 23-m (75-ft) high pier in order to study the influence of bridge height on the results. In this case, only the wind data from the St. Louis site are used.

The results of the calibration are summarized in Table 3.19. As an example, the calibration process follows this logic: if the 7.6-m (25-ft) bridge column is designed to satisfy the AASHTO LRFD specifications, then the required moment capacity for the column will be $M_{cap} = M_{design} = 2.5$ MN-m (1,835 kip-ft). Choosing $M_{design} = 2.5$ MN-m (1,835 kip-ft) for the Austin wind data will produce a reliability index (for a 75-year bridge design life) equal to $\beta = 3.92$ if no live load is considered. When the bridge column is designed for live load, then the required moment capacity is $M_{cap} = M_{design} = 8$ MN-m (5,917 kip-ft), producing a reliability index $\beta = 3.71$. If one considers that live loads may occur within the 4-h period when a windstorm is acting on the bridge column as well as when no wind is on, then the reliability index of the column is reduced as shown in Figure 3.28. To increase the reliability index back to the original $\beta = 3.71$ obtained for gravity loads alone, the moment capacity of the column should be increased. By interpolation, the moment capacity that will produce a reliability index equal to $\beta = 3.71$ when both wind loads and live loads are considered should be $M_{required} = 8.2$ MN-m (6,038 kip-ft) or an increase of 200 kN-m (121 = 6,038 − 5,917 kip-ft).

For the case in which wind alone is applied, the design moment capacity (M_{design}) value is 2.5 MN-m (1,835 kip-ft), which includes the effect of the resistance factor for bending of $\phi = 0.90$ and a wind load factor of $\gamma_{WS} = 1.4$ (see Equation

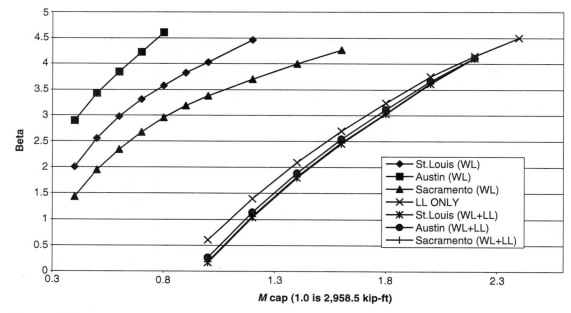

Reliability index for one column under wind load and live load (concrete)

Figure 3.28. Reliability index for combination of wind loads and live load for moment capacity limit state.

83

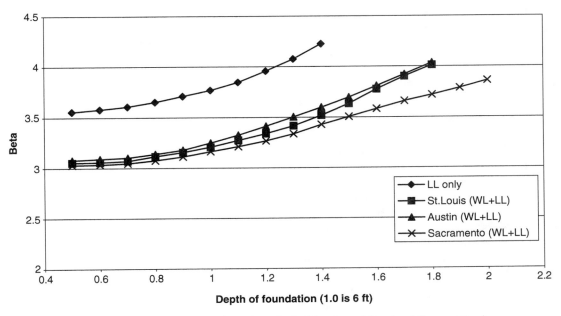

Reliability index for overtipping of one column under wind load and live load

Figure 3.29. Reliability index for combination of wind loads and live load for overtipping.

3.4 and Table 3.2). The moment calculated from the wind load is equal to $M_{WS} = 1.6$ MN-m (1,180 kip-ft) for a design wind speed of 145 km/h (90 mph). Also, when the live load is on the structure, the AASHTO LRFD requires the consideration of the wind load on live load. For a 90 km/h (56 mph)

wind speed, the AASHTO LRFD recommends applying a distributed force of 1.46 kN/m (0.1 kip/ft). This would result in a moment at the base of the column equal to $M_{WL} = 513$ kN-m (378 kip-ft). Finally, the applied live load from the HL-93 live loading produces a moment in the column equal

TABLE 3.19 Summary of live load factors for combination of *WS* + *LL* for moment capacity

Site	M_{req} (kip-ft)	M_{WS}	M_{WL}	M_{LL}	γ_{LL}	$\gamma_{WS} = \gamma_{WL}$
St. Louis (25-ft)	6,086	1,180	378	3,048	1.35	0.87
Austin (25-ft)	6,038	1,180	378	3,048	1.35	0.85
Sacramento (25-ft)	6,074	1,180	378	3,048	1.35	0.87
St. Louis (75-ft)	13,532	4,927	2,673	3,048	1.35	1.10
Average						*0.92*
St. Louis (25-ft)	6,086	1,180	378	3,048	1.50	0.58
Austin (25-ft)	6,038	1,180	378	3,048	1.50	0.55
Sacramento (25-ft)	6,074	1,180	378	3,048	1.50	0.57
St. Louis (75-ft)	13,532	4,927	2,673	3,048	1.50	1.00
Average						*0.68*
St. Louis (25-ft)	6,086	1,180	378	3,048	1.75	0.09
Austin (25-ft)	6,038	1,180	378	3,048	1.75	0.06
Sacramento (25-ft)	6,074	1,180	378	3,048	1.75	0.09
St. Louis (75-ft)	13,532	4,927	2,673	3,048	1.75	0.90
Average						*0.47*

to $M_{LL} = 4.1$ MN-m (3,048 kip-ft). The live load factor that should be used for the combination of earthquakes and live load should be determined such that

$$\phi\, M_{req} = \gamma_{WS}\, M_{WS} + \gamma_{WL}\, M_{WL} + \gamma_{LL}\, M_{LL} \qquad (3.35)$$

or

$$0.90\, M_{req} = \gamma_{WS}\, 1{,}600 \text{ kN-m} + \gamma_{WL}\, 512 \text{ kN-m} + \gamma_{LL}\, 4132 \text{ kN-m}.$$

The calibration of the load combination factors involves the determination of the values of γ_{WS}, γ_{WL} and γ_{LL} needed to produce the required column moment capacity $M_{req} = 8.1$ MN-m (6,038 kip-ft). Following the format of the current AASHTO LRFD specifications, it is suggested that γ_{WS} be set equal to γ_{WL}. Several options for $\gamma_{WS} = \gamma_{WL}$ and γ_{LL} are possible to match the required moment capacity. Table 3.19 presents some of these options. For example, if the live load factor is chosen such that $\gamma_{LL} = 1.75$, then the corresponding values for the wind load is given as $\gamma_{WS} = \gamma_{WL} = 0.06$ for the Austin data. If $\gamma_{LL} = 1.35$ is selected as is the case in the current AASHTO LRFD, then the required wind load factor becomes $\gamma_{WS} = \gamma_{WL} = 0.85$ for the Austin wind data. Similar calculations are executed for all the three wind site data selected in this section—St. Louis, Austin, and Sacramento—and for two bridge column heights—7.6 m (25 ft) and 23 m (75 ft)—for column failure in bending. The results are summarized in Table 3.19. Similar calculations performed for column overtipping are summarized in Table 3.20.

For overtipping of short piles, the calibration is based on the results shown in Figure 3.29. As an example, the calibration is executed as follows: the required foundation depth assuming the St. Louis wind data should be $L_{req} = 2.9$ m (9.57 ft). This depth is achieved by balancing the following equation:

$$\phi_{soil} \frac{3\gamma D K_p L_{req}^3}{3x2} + \phi_{DC} DC\left(\frac{D}{2}\right) = \gamma_{WS} F_{WS1}(8.8 + L_{req})$$
$$+ \gamma_{WS} F_{WS2}(3.6 + L_{req}) \qquad (3.36)$$
$$+ \gamma_{WL} F_{WL}(12.1 + L_{req})$$
$$+ \gamma_{LL} M_{LL}$$

where

DC = the dead weight of 6.5 MN (1,450 kips) applied at the center of the column with a counteracting load factor of $\phi_{DC} = 0.90$,
$\phi_{soil} = 0.50$ = the soil's resistance factor,
γ = the specific weight of the soil,
D = the column diameter,
K_p = the Rankine coefficient,
L_{req} = the depth of the foundation,
M_{LL} = the applied unfactored live moment where $M_{LL} = 3.1$ MN-m (2,286 kip-ft).

A different live load moment is used for overtipping as compared with bending because for overtipping, the moment is taken about the edge of the column, not the center. The wind force on the structure is $F_{WS1} = 141$ kN (31.6 kips) applied at 8.8 m (29 ft) above the soil, the wind force on the column is $F_{WS2} = 27$ kN (6 kips) applied at 3.6 m (11.8 ft), and the wind force on the live load is $F_{WL} = 42.33$ kN (9.5 kips) applied at 12 m (39.7 ft). Equation 3.36 becomes

TABLE 3.20 Summary of live load factors for combination of WS + LL for overtipping

Site	L_{req} (ft)	F_{WS1} (kip)	F_{WS2} (kip)	F_{WL} (kip)	M_{LL} (kip-ft)	γ_{LL}	$\gamma_{WS} = \gamma_{WL}$
St. Louis (25-ft)	9.6	31.6	6	9.5	2,286	1.35	0.62
Austin (25-ft)	9.4	31.6	6	9.5	2,286	1.35	0.61
Sacramento (25-ft)	11.2	31.6	6	9.5	2,286	1.35	0.68
Average							*0.64*
St. Louis (25-ft)	9.6	31.6	6	9.5	2,286	1.50	0.43
Austin (25-ft)	9.4	31.6	6	9.5	2,286	1.50	0.42
Sacramento (25-ft)	11.2	31.6	6	9.5	2,286	1.50	0.50
Average							*0.45*
St. Louis (25-ft)	9.6	31.6	6	9.5	2,286	1.75	0.11
Austin (25-ft)	9.4	31.6	6	9.5	2,286	1.75	0.10
Sacramento (25-ft)	11.2	31.6	6	9.5	2,286	1.75	0.20
Average							*0.14*

$$0.5 \frac{3 * 9.4 kN/m^3 * 1.8 * 3.68 L_{\text{req}}^3}{3 x 2} + 0.9 * 6.5 * 10^3 \left(\frac{1.8}{2}\right)$$
$$= \gamma_{WS} 141(8.8 + L_{\text{req}}) + \gamma_{WS} 26.7(3.6 + L_{\text{req}}) \qquad (3.36')$$
$$+ \gamma_{WL} 42.3(12.1 + L_{\text{req}}) + \gamma_{LL} 3.1 * 10^3$$

given that $L_{\text{req}} = 2.9$ m (9.57 ft) and solving for $\gamma_{WS} = \gamma_{WL}$, given a value of γ_{WL} will provide the appropriate load combination factors to be considered. These are provided in Table 3.20. Differences are observed between the results for overtipping and bending moment failures and for different column heights in Tables 3.19 and 3.20. These differences are mainly due to the different influences of the statistical uncertainties associated with modeling the soil resistance and column bending capacity.

The results of Tables 3.19 and 3.20 show that the lower live load factors, γ_{LL}, would provide more uniform ranges of $\gamma_{WS} = \gamma_{WL}$ values. In particular, if the current AASHTO recommended value of $\gamma_{LL} = 1.35$ is used, then the corresponding average wind load factor set obtained from overtipping and column bending is $\gamma_{WS} = \gamma_{WL} = 0.80$. This 0.80 value is higher than the current AASHTO LRFD set of $\gamma_{WS} = \gamma_{WL} = 0.40$, indicating that the current code produces relatively low safety levels. An optimization algorithm is also used to determine the load combination factors that would minimize the sum of the square of the differences between the required moment capacities and those that would be obtained from the right hand side of Equations 3.35 and 3.36. The optimization algorithm indicates that the differences are minimized when the combination of $\gamma_{WS} = \gamma_{WL} = 1.17$ and $\gamma_{LL} = 1.06$ is used. Based on these results, it is herein recommended that the combination $\gamma_{WS} = \gamma_{WL} = 1.20$ and $\gamma_{LL} = 1.00$ be used for bridges subjected to the combined threats of wind loads and live loads.

3.3.3 Combination of Scour and Live Loads (SC + LL)

In this section, the single-column bent described in Figure 3.1 is analyzed to illustrate how the combined effects of scour and live loads will affect the bent's reliability. Scour data from three river sites—the Schohaire, Sandusky, and Rocky—are selected as representative data. Table 2.8 of Chapter 2 provides the basic river discharge data for the selected sites. The live load data are obtained from the models developed by Nowak (1999) for NCHRP Project 12-33 as described in 2.4.1 (Table 2.3). The reliability calculations follow the Ferry-Borges model described in Section 2.5. During the calculations, the following assumptions are made:

- Scour depths last for 6 months, during which time the erosion of the soil around the column base remains at its highest value. The scour depth remains constant throughout the 6-month period.

- The scour depths are independent from year to year.
- Each site is on the average exposed to one major scour in 1 year.
- The probability distribution of the maximum yearly discharge rate follows a lognormal distribution. This maximum yearly discharge rate is assumed to control the scour depth for the year.
- The live load model assumes that for each one-lane loading event, the live load effect is on the average 0.79 times the effect of the HL-93 loading configuration with a COV of 10%. For two lanes, the average event produces an average load effect equal to1.58 times the effect of one lane of HL-93 with a COV of 7%.
- Each site will be exposed to 1,000 single-lane heavy truck and 67 two-lane truck events in 1 day based on the assumptions of Nowak (1999) and Moses (2001).
- Equation 2.11 is used to find the probability distribution of the live load for different return periods. In particular, the live load for a 6-month period is calculated for studying the reliability caused by the application of live loads when scour is on.
- The effect of the scour depth in reducing the soil resistance around the column base is combined with the effects of the 6-month live load magnitude and projected to provide the maximum expected combined load in the 75-year bridge design life period using the Ferry-Borges model described in Section 2.5. The calculations account for the cases in which the live load occurs with the 6-month scour on and the cases in which the live load occurs with no scour on.
- The reliability calculations are executed for column overtipping because the bending moment in deeply embedded columns is not affected by the presence of scour.
- The reliability calculations account for the uncertainties associated with predicting the discharge rate, estimating the scour depth given a discharge rate, projecting the live load magnitude, estimating the counter effect because of the dead loads, and estimating the soil resistance properties.
- Referring to Figure 3.4(a), the failure equation for overtipping may be represented as follows:

$$Z = 0 - \max \left\{ \begin{array}{l} \max_{n_1} \left[\frac{3\gamma D K_p (L - y_{sc})^3}{6} + F_{DC}\frac{D}{2} - F_{LL}^*\left(e_3 - \frac{D}{2}\right) \right] \\ \max_{n_2} \left[\frac{3\gamma D K_p (L)^3}{6} + F_{DC}\frac{D}{2} - F_{LL}^{**}\left(e_3 - \frac{D}{2}\right) \right] \end{array} \right\} \quad (3.37)$$

where

F_{DC} = the dead weight applied on the column,
F_{LL}^* = the live load that occurs when scour is on,
F_{LL}^{**} = the live load when there is no scour,
γ = the specific weight of the soil,
D = the column diameter,

K_p = the Rankine coefficient,
L = the foundation depth,
e_3 = the eccentricity of the live load relative to the center of the column, and
y_{sc} = the scour depth.

In this case, $F_{LL}*$ is the maximum live load effect that may occur within the 6-month period when scour occurs. n_1 represents the number of scours expected in a 75-year period ($n_1 = 75$). n_2 is the number of live load events expected when there is no scour (e.g., $n_2 = 75 * 6 * 30 * 1{,}000$ for one-lane loading events).

The results of the reliability analysis are provided in Figure 3.30 for overtipping of the single-column bent. The data are presented for the three representative river sites for combination of scour and live loads and for live loads alone. The results are also summarized in Table 3.21. As an example, if the 7.6-m (25-ft) bridge column is designed to resist overtipping caused by the application of the HL-93 live load of the AASHTO LRFD specifications, then the column should be embedded a distance $L = 1.8$ m (6 ft) into the soil. Choosing $L = 1.8$ m (6 ft) with no scour would produce a reliability index of $\beta = 3.68$. If scour occurs because of a river having the discharge rate of the Rocky River, then the reliability index reduces to $\beta = -2.0$ because of the possible combination of live load and scour in the 75-year bridge design life of a bridge column with a foundation depth $L = 1.8$ m. However, if one wishes to increase the reliability index from -2.0 back to the original 3.68, then the depth of the column should be increased. By interpolation, the required foundation depth that will produce a reliability index equal to $\beta = 3.68$ when both scour and live loads are considered should be $L_{req} = 7.4$ m (24.4 ft) or an increase by a factor of 4.1. Table 3.4 shows

that the expected scour depth from the Rocky River is 3.7 m (12.3 ft). Hence, to go from $L = 1.8$ m to L = 7.4 m (6 ft to 24.4 ft), one needs to include 1.50 times the design scour depth. Thus, the load factor for scour would be equal to $\gamma_{SC} = 1.50$. In other words, for determining the required design scour for the combination of scour plus live loads, one should use the scour HEC-18 equation (Equation 2.22) then multiply the value obtained by a load factor equal to 1.50.

Similar calculations are executed for all three scour site data selected in this section: the Schohaire, Sandusky, and Rocky Rivers. The results for all these cases are summarized in Table 3.21. The average scour factor for the three cases analyzed is $\gamma_{SC} = 1.79$, which indicates that a scour factor of $\gamma_{SC} = 1.80$ associated with a live load factor of $\gamma_{LL} = 1.75$ is reasonable when studying the combination of scour and live loads.

The recommended scour load factor $\gamma_{SC} = 1.80$ reflects the fact that the current scour model provides a low reliability level compared with that for the live loads. As seen above, the scour alone model gives an average reliability index close to 1.0 as compared with $\beta = 3.50$ to 3.70 for live loads. If the goal is to have bridges with foundations set in water channels satisfy the same safety requirements as bridges not subject to scour, then a live load factor $\gamma_{LL} = 1.75$ should be used in combination with a scour factor $\gamma_{SC} = 1.80$. One should note that even for multicolumn bents, failure caused by scour will generally result in the collapse of the system as the bent system would provide little redundancy when one column loses its ability to carry load in a sudden (brittle) failure. Hence, using a system factor of 0.80 in addition to the member resistance factor would be appropriate during the design of column bents for combination of scour and other loads. The application of system factors during the design and safety

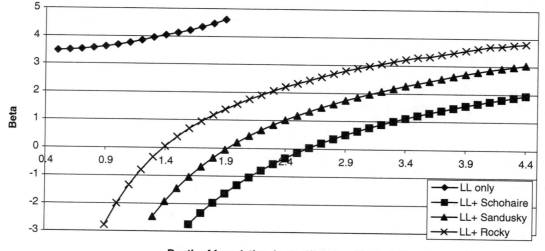

Figure 3.30. Reliability index for combination of scour and live load.

TABLE 3.21 Summary of live load factors for combination of *SC* plus *LL*

Site	L_{req} (ft)	L_{SC}	L_{LL}	γ_{SC} for *LL* + *SC*
Schohaire	41.0 ft	17.3 ft	6 ft	2.02
Sandusky	32.3 ft	14.3 ft	6 ft	1.84
Rocky	24.4 ft	12.3 ft	6 ft	1.50
Average				*1.79*

evaluation of bridge bents has been discussed elsewhere by Liu et al. (2001).

3.3.4 Combination of Scour and Wind Load (*SC* + *WL*)

The single-column bent described in Figure 3.1 is analyzed to illustrate how the combined effects of scour and wind loads will affect the bent's reliability. Scour data from three river sites—the Schohaire, Sandusky, and Rocky—are selected as representative sites. Table 2.8 of Chapter 2 provides the basic river discharge data for the selected sites. In addition, wind speed data from Austin, Sacramento, and St. Louis are selected as representatives of wind speed data. The reliability calculations follow the Ferry-Borges model described in Section 2.5. The following assumptions are made:

- Scour depths last for 6 months, during which time the erosion of the soil around the column base remains at its highest value. The scour depth remains constant throughout the 6-month period.
- Scour depths are independent from year to year.
- Each site is on the average exposed to one major scour in 1 year.
- The probability distribution of the maximum yearly discharge rate follows a lognormal distribution. This maximum yearly discharge rate is assumed to control the scour depth for the year.
- The wind speed data are assumed to follow a Gumbel distribution with mean and COV values as shown in Table 2.7.
- Each site will be exposed to 200 winds in a year.
- Equation 2.14 is used to find the probability distribution of the wind load for a 6-month period. This is used for studying the reliability caused by the application of wind loads when scour is on.
- The effect of the scour depth in increasing the moment arm of the wind load is combined with the effects of the 6-month wind load and projected to provide the maximum expected combined load in the 75-year bridge design life period using the Ferry-Borges model described in Section 2.5. The calculations account for the cases in which the wind load occurs with the 6-month scour on and the cases in which the wind occurs when no scour is on.

- The reliability calculations are executed for column bending because overtipping is mostly balanced by the weight of the structure and is not significantly affected by the presence of scour.
- The reliability calculations account for the uncertainties associated with predicting the discharge rate, estimating the scour depth given a discharge rate, projecting the wind load magnitude, estimating the counter effect caused by the soil resistance, and estimating the moment capacity of the column.
- Referring to Figure 3.4, the failure equation for bending in the column may be represented as follows:

$$Z = M_{col} + \frac{3\gamma D K_p (f)^3}{6}\lambda_{cyc} - \max\left\{\begin{array}{l} \max_{n_1}\left[\left(\max_{n_2} F_{WS}^*\right)(e_1 + y_{sc} + f)\right] \\ \max_{n_3}\left[F_{WS}^{**}(e_1 + f)\right] \end{array}\right\} \quad (3.38)$$

where

M_{col} = the bending moment capacity of the column;
F_{WS}^* = the resultant wind load on the structure that occurs when scour is on (it accounts for wind on superstructure and wind on column);
F_{WS}^{**} = the resultant wind load on the structure when there is no scour;
γ = the specific weight of the soil;
D = the column diameter;
K_p = the Rankine coefficient;
f = the distance below the soil level at which the maximum bending moment occurs;
λ_{cyc} = the factor that accounts for the effect of cyclic loads on foundation strength;
e_1 = the eccentricity of the resultant wind load relative to the soil level (before scour);
y_{sc} = the scour depth;
n_1 = the number of scours expected in a 75-year period ($n_1 = 75$);
n_2 = the number of wind loads expected when there is scour ($n_2 = 100$); and
n_3 = the number of wind loads that occur during the design life of the bridge when there is no scour ($n_3 = 75 * 100$).

The results of the reliability analysis are provided in Figure 3.31 for bending of the single-column bent. The data are presented for the three representative river sites for combination of scour and three different wind loads and for wind loads alone. The results are also summarized in Table 3.22. As an example, if the 8-m (25-ft) bridge column is designed to resist failure in bending caused by the application of the AASHTO design wind load, then the column should have a design moment capacity of $M_{design} = 2.5$ MN-m (1,835 kip-ft). If the bridge is situated in a region in which the wind data are similar to that observed in Austin, then without the application of any live load the reliability index would be $\beta = 3.92$. If scour occurs because of a river having the discharge rate of the Schohaire, then the reliability index reduces to $\beta = 3.63$ because of the possible combination of wind load and scour in the 75-year bridge design life when the column has a bending capacity of $M_{cap} = M_{design} = 2.5$ MN-m (1,835 kip-ft). However, if one wishes to increase the reliability index from 3.63 back to the original 3.92, then the column capacity should be increased. By interpolation, the required moment capacity that will produce a reliability index equal to $\beta = 3.92$ when both scour and wind loads are considered should be $M_{cap} = 2.9$ MN-m (2,139 kip-ft) or an increase by a factor of 1.15. The scour depth that should be included to achieve this required M_{cap} is calculated from Equation 3.39:

$$\phi M_{req} = \gamma_{WS} M_{WS} \qquad (3.39)$$

or

$$0.90 M_{req} = \gamma_{WS} F_{WS1} (8.8 \text{ m} + \gamma_{SC} y_{sc})$$
$$+ \gamma_{WS} F_{WS2} (3.6 \text{ m} + \gamma_{SC} y_{sc})$$

where

M_{WS} = the moment exerted on the column because of the applied design wind pressure when the moment arm of the force has been extended by the required factored scour depth;

$F_{WS1} = 140$ kN (31.6 kips), the wind design force on the superstructure located at a distance of 8.8 m (29 ft) above the original soil level;

$F_{WS2} = 27$ kN (6 kips), the wind design force on the column located at a distance of 3.6 m (11.8 ft) above the original soil level.

Given that the scour equation produces $y_{sc} = 5.3$ m (17.3 ft) for the Schohaire River data and using a wind factor $\gamma_{WS} = 1.40$ as stipulated by the AASHTO LRFD for wind on structures, the required scour factor is calculated to be $\gamma_{SC} = 0.60$.

Similar calculations are executed for all three wind site data (St. Louis, Austin, and Sacramento) and for the three scour site data selected in this section (the Schohaire, Sandusky, and Rocky Rivers). The results for all these cases are summarized in Table 3.22. It is noticed that the range of values for γ_{SC} is between a low of 0.53 and a high of 0.91. The reason for this spread in the results is due to the large range in the reliability index associated with the wind loads and the scour depths. This observation confirms the need to develop new wind maps that would better reflect the actual variations in the wind speeds at different sites in the United States as well as the development of new scour design equations that would reduce the observed differences for rivers with different discharge rates.

From the results shown in Table 3.22, it seems appropriate to conclude that a load factor on scour equal to $\gamma_{SC} = 0.70$

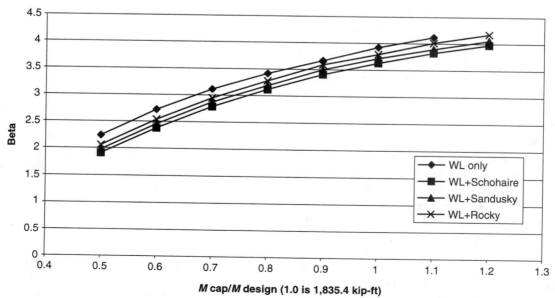

Reliability index for one column under wind load and scour (*WL*: Austin, TX)

Figure 3.31. Reliability index for combination of wind loads plus scour.

TABLE 3.22 Summary of live load factors for combination of *SC* plus *WS*

Wind data	Scour data	M_{req}	F_{WS1}	F_{WS2}	y_{sc} (ft)	γ_{SC} for *WS + SC*
Austin	Schohaire	2,139 (kip-ft)	31.6 (kips)	6 (kips)	17.3	0.60
Austin	Sandusky	2,050	31.6	6	14.3	0.62
Austin	Rocky	1,945	31.6	6	12.3	0.57
Sacramento	Schohaire	2,314	31.6	6	17.3	0.77
Sacramento	Sandusky	2,260	31.6	6	14.3	0.87
Sacramento	Rocky	2,191	31.6	6	12.3	0.91
St. Louis	Schohaire	2,072	31.6	6	17.3	0.53
St. Louis	Sandusky	2,004	31.6	6	14.3	0.56
St. Louis	Rocky	1,959	31.6	6	12.3	0.59
Average						*0.67*

in combination with a wind load factor of $\gamma_{WS} = 1.40$ is appropriate for use when studying the combination of scour and wind loads. This $\gamma_{SC} = 0.70$ value is lower than that used for combining scour and live loads because there are fewer wind storms expected in the 75-year design life of a bridge structure as compared with the number of truck loading events. Fewer load occurrences imply a lower chance for combining high wind load intensity with high scour depth.

3.3.5 Combination of Scour and Vessel Collision (*SC + CV*)

The single-column bent described in Figure 3.1 is analyzed to illustrate how the combined effects of scour and vessel collision forces will affect the bent's reliability. Scour data from the Mississippi River at the location of the I-40 bridge are used in combination with the barge traffic at that location. Appendix C provides the basic river discharge data for the selected site. In addition, vessel collision forces as assembled in Figure 2.13 are used. The reliability calculations follow the Ferry-Borges model described in Section 2.5. The following assumptions are made:

- Scour depths last for 6 months, during which time the erosion of the soil around the column base remains at its highest value. The scour depth remains constant throughout the 6-month period.
- The scour depths are independent from year to year.
- The site is on the average exposed to one major scour in 1 year.
- The probability distribution of the maximum yearly discharge rate follows a lognormal distribution. This maximum yearly discharge rate is assumed to control the scour depth for the year. It is noted that because Equation 2.31 was developed based on data from small rivers, this model may not be valid for rivers with high discharge rates. For this reason, this analysis is based on the model developed in Appendix B based on the data provided by Johnson and Dock (1998) as expressed in

Equation 2.30 with a scour modeling variable, λ_{sc}, having a mean value equal to 0.55 and a COV of 52%.
- The collision force data are assumed to follow the probability distribution curve shown in Figure 2.13.
- Each site will be exposed to 0.83 barge flotilla collisions in 1 year.
- Equation 2.11 is used to find the probability distribution of the wind load for a 6-month period for studying the reliability caused by the application of wind loads when scour is on.
- The effect of the scour depth in increasing the moment arm of the collision force is combined with the effects of the 6-month collision force and projected to provide the maximum expected combined load in the 75-year bridge design life period using the Ferry-Borges model described in Section 2.5. The calculations account for the cases in which the collision occurs with the 6-month scour on and the cases in which the collision occurs when no scour is on.
- The reliability calculations are executed for column bending because overtipping is mostly balanced by the weight of the structure and is not significantly affected by the presence of scour.
- The reliability calculations account for the uncertainties associated with predicting the discharge rate, estimating the scour depth given a discharge rate, projecting the collision force magnitude, estimating the counter effect caused by the soil resistance, and estimating the moment capacity of the column.
- Referring to Figure 3.4, the failure equation for bending in the column may be represented as follows:

$$Z = M_{col} + \frac{3\gamma D K_p (f)^3}{6}\lambda_{cyc}$$
$$- \max\left\{\begin{array}{l} \max_{n_1}\left[\left(\max_{n_2} F_{CV}^*\right)(e_1 + y_{sc} + f)\right] \\ \max_{n_3}\left[F_{CV}^{**}(e_1 + f)\right] \end{array}\right\} \quad (3.40)$$

where

M_{col} = the bending moment capacity of the column,
F_{CV}^* = the collision force on the structure that occurs when scour is on,
F_{CV}^{**} = the collision force on the structure when there is no scour,
γ = the specific weight of the soil,
D = the column diameter,
K_p = the Rankine coefficient,
f = the distance below the soil level at which the maximum bending moment occurs,
λ_{cyc} = the factor that accounts for the effect of cyclic loads on foundation strength,
e_1 = the eccentricity of the collision force relative to the soil level (before scour),
y_{sc} = the scour depth,
n_1 = the number of scours expected in a 75-year period ($n_1 = 75$),
n_2 = the number of collisions expected within 1 year when there is scour ($n_2 = 0.42$), and
n_3 = the number of collisions that occur when there is no scour ($n_3 = 0.42 * 75$).

The results of the reliability analysis are provided in Figure 3.32 for bending of the single-column bent. The results show that if the 150-ft bridge column is designed to resist failure in bending caused by the application of the collision forces determined using the AASHTO specifications and following the model of Whitney et al. (1996), the column should

have a design moment capacity $M_{design} = 464$ MN-m (342,333 kip-ft). The corresponding reliability index—assuming that no scour is possible—would be $\beta = 2.78$. If the bridge is exposed to scour because of a river having the discharge rate of the Mississippi and the channel profile shown in Figure 3.6, then the reliability index reduces to $\beta = 2.05$ because of the possible combination of collision forces and scour in the 75-year bridge design life when the column capacity of $M_{cap} = M_{design} = 464$ MN-m (342,333 kip-ft). However, if we wish to increase the reliability index from 2.05 back to the original 2.78, then the column capacity should be increased. By interpolation, the required moment capacity that will produce a reliability index equal to $\beta = 2.78$ when both scour and vessel collisions are considered should be $M_{cap} = 715$ MN-m (527,200 kip-ft) or an increase by a factor of 1.54. To achieve this M_{cap}, the following design equation should be satisfied:

$$\phi M_{req} = \gamma_{CV} M_{CV} \tag{3.41}$$

or

$$0.90 M_{req} = \gamma_{CV} F_{CV} (4.9m + \gamma_{SC} y_{SC} + f)$$
$$- \frac{3\gamma D K_p f^3}{6}$$

where

M_{CV} = the moment exerted on the column caused by the applied design wind pressure when the moment arm

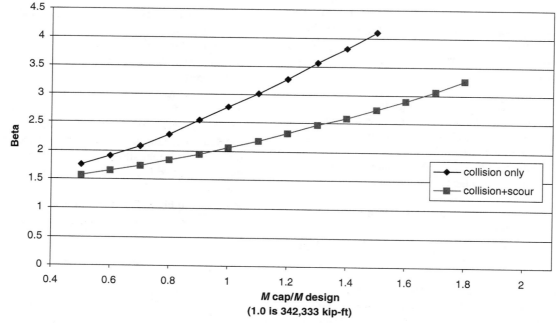

Figure 3.32.   Reliability index for combination of collision and scour.

of the force has been extended by the required factored scour depth;

γ = the specific weight of the soil;

D = the column diameter;

K_p = the Rankine coefficient;

F_{CV} = 35 MN (7,900 kips), the vessel collision design force on the column applied at a distance of 16 ft above the original soil level; and

f = 10.5 m (34.5 ft) = the depth to the point of maximum moment.

Note that in this equation the effect of the soil resistance is included because f is reasonably deep. For the cases analyzed above (e.g., wind loads), f is relatively small and the contributions of the soil resistance does not significantly affect the design capacity of the column. Given that the scour equation produces y_{sc} = 10.4 m (34.2 ft) for the Mississippi River data and using a vessel collision factor $\gamma_{CV} = 1.0$ as stipulated by the AASHTO LRFD for collision forces, the required scour factor is calculated to be $\gamma_{SC} = 0.62$.

From the calculations shown above, it seems appropriate to conclude that a load factor on scour equal to $\gamma_{SC} = 0.60$ associated with $\gamma_{CV} = 1.0$ is reasonable for use when studying the combination of scour and ship collision forces. This value is lower than that used for combining scour and live loads and for combining wind and scour because there are fewer collisions expected in the 75-year design life of a bridge structure as compared with the number of truck loading events or windstorms. Fewer load occurrences imply a lower chance for combining high vessel collision forces with high scour depths. Also, the target reliability index selected for the calibration is lower than that used for live loads and is lower than that obtained from the average of different wind data. The lower target reliability index is selected to match the reliability index observed for the cases in which vessel collisions occur without risk of scour.

3.3.6 Combination of Vessel Collision and Wind Loads (*CV* + *WS*)

In this section, the single-column bent described in Figure 3.1 is analyzed to illustrate how the combined effects of vessel collision forces and wind loads will affect its reliability. It is herein assumed that wind load data collected in Knoxville, Tennessee, are applicable for this bridge site. This Knoxville wind data are used in combination with the barge traffic at the location of the I-40 bridge in Memphis. In addition, vessel collision forces as assembled in Figure 2.13 are used. The reliability calculations follow the Ferry-Borges model described in Section 2.5. The following assumptions are made:

- The wind speed data are assumed to follow a Gumbel distribution, with a yearly mean value equal to 77.8 km/h (48.6 mph) and COV of 14%.
- The site is exposed to 200 winds in 1 year.

- Equation 2.14 is used to find the probability distribution of the wind load for a 6-month period for studying the reliability caused by the application of wind loads when scour is on.
- The collision force data are assumed to follow the probability distribution curve shown in Figure 2.13.
- Each site will be exposed to an average of 0.83 barge flotilla collisions in 1 year.
- Equation 2.14 is used to find the probability distribution of the wind load for a 6-month period for studying the reliability caused by the application of a collision force when the wind is on.
- A correlation between the wind speed and the number of collisions is assumed based on the data provided by Larsen (1993). According to this model, the rate of collisions increases by a factor of 3 (from 0.83 to 2.50 collisions/year) when the wind speed increases from 6 m/sec (13 mph) to 14 m/sec (31 mph). Below 6 m/sec (13 mph), the number of collisions remains at 0.83 collisions/year; beyond 14 m/sec (31 mph), the number of collision remains at a rate of 2.50/year. Although barges and flotilla may not travel when the wind speeds exceed the 14 m/sec (31 mph) limit, it is herein assumed that vessels and barges may break from their moorings when there is a large windstorm such that collisions are still possible.
- The effect of the moment from wind forces is combined with the effects of the collision force and projected to provide the maximum expected combined load in the 75-year bridge design life period using the Ferry-Borges model described in Section 2.5. The calculations account for the cases in which the collision occurs during a windstorm and the cases in which the collision occurs when no wind is on.
- The reliability calculations are executed for column bending because overtipping is mostly balanced by the weight of the structure.
- The reliability calculations account for the uncertainties associated with predicting the wind speed, estimating the wind forces, projecting the collision force magnitude, estimating the counter effect caused by the soil resistance, and estimating the moment capacity of the column.
- Referring to Figure 3.4, the failure equation for bending in the column may be represented as follows:

$$Z = M_{col} + \frac{3\gamma DK_p(f)^3}{6}\lambda_{cyc}$$
$$- \max\left\{ \begin{array}{c} \max_{n_1}\left[F_{WS}(e_2 + f) + \left(\max_{n_2} F_{CV}^*\right)(e_2 + f)\right] \\ \max_{n_3}\left[F_{CV}^{**}(e_2 + f)\right] \end{array} \right\} \quad (3.42)$$

where

M_{col} = the bending moment capacity of the column,

$F_{CV}*$ = the collision force on the structure that occurs when a wind is on,

$F_{CV}**$ = the collision force on the structure when there is no wind,

γ = the specific weight of the soil,

D = the column diameter,

K_p = the Rankine coefficient,

f = the distance below the soil level at which the maximum bending moment occurs,

λ_{cyc} = the factor that accounts for the effect of cyclic loads on foundation strength,

e_2 = the eccentricity of the collision force relative to the soil level,

e_1 = the moment arm of the resultant wind force,

n_1 = the number of winds expected in a 75-year period ($n_1 = 200 \times 75$),

n_2 = the number of collisions expected when there is a wind, and

n_3 = the number of collisions that occur when there is no wind.

In order to study the effects of the winds on the structure and calculate the wind force, F_{WS}, the results provided in Appendix F are used. The results essentially show that at a wind of 70 mph the wind force is equal to 18.6 MN (4,180 kips) and that the resultant force will be applied at an eccentricity $e_1 =$ 68 m (223 ft) from the soil's surface. Interpolation as a func-

tion of V^2 (where V = wind velocity) is used to find the wind forces for different wind speeds. Notice that the force eccentricity is larger than the column height because of the effect of the superstructure.

The results of the reliability analysis are provided in Figure 3.33 for bending of the single-column bent. For example, if the 46-m (150-ft) bridge column is designed to resist failure in bending caused by the application of the collision forces determined using the AASHTO specifications and following the model of Whitney et al. (1996), then the column should have a design moment capacity of $M_{design} = 464$ MN-m (342,333 kip-ft). If the bridge was originally designed without taking into consideration the possibility of windstorms, the reliability index would be $\beta = 2.78$. If the bridge is exposed to winds in addition to collision forces, then the reliability index reduces to $\beta = 0.82$ in the 75-year bridge design life when the column has a capacity $M_{cap} = M_{design} = 464$ MN-m (342,333 kip-ft). However, if one wishes to increase the reliability index from 0.82 back to the original 2.78, then the column capacity should be increased. By interpolation, the required moment capacity that will produce a reliability index equal to $\beta = 2.78$ when both vessel collision forces and wind loads are considered should be $M_{cap} = 896$ MN-m (660,700 kip-ft) or an increase by a factor of 1.93. To achieve this M_{cap}, load combination factors should be used in the design equations. The determination of the appropriate load factors is executed using the following equation:

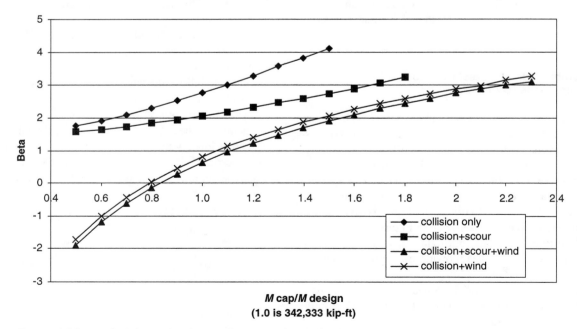

Figure 3.33. Reliability index for combination of vessel collision plus wind loads and scour.

$$\phi M_{req} = \gamma_{WS} M_{WS} + \gamma_{CV} M_{CV} \qquad (3.43)$$

or

$$0.90 M_{req} = \gamma_{WS} 18.6 \text{ MN} * (68 \text{ m} + 2f/3) \\ + \gamma_{CV} 35 \text{ MN} * (4.9 \text{ m} + 2f/3).$$

By trial and error, it has been determined that a $\gamma_{WS} = 0.27$ associated with $\gamma_{CV} = 1.0$ would lead to a point of maximum moment located at a distance of $f = 11.3$ m (37 ft) below the soil level. These factors will produce a required moment capacity $M_{req} = 896$ MN-m (660,700 kip-ft), which will lead to a reliability index of $\beta = 2.78$ for the combination of wind loads and vessel collisions. The $\beta = 2.78$ is the same reliability index obtained from vessel collision forces alone.

In addition to showing how the reliability index varies for the cases in which vessel collision forces are applied alone on the bridge, Figure 3.33 shows how the combination of wind forces, collision forces, and scour affect the reliability index. It is noted that the probability of having all three extreme events simultaneously is very small. Hence, the drop in the reliability index from the case in which wind and vessel collision forces are combined to the case in which wind, scour, and vessel collision are combined is less than 0.1. Therefore, it is not necessary to check this possible combination. Based on the calculations performed in this section it is recommended to use a wind load factor of $\gamma_{WS} = 0.30$ in combination with $\gamma_{CV} = 1.0$ when checking the safety of bridges under the combined effects of winds and vessel collisions. The 0.30 wind factor proposed is lower than the 0.80 factor used for combinations involving live loads. This is justified based on the lower number of vessel collisions expected when compared with the number of live load events, meaning there is a lower chance of combination even though a correlation between the rate of collisions is made with the intensity of the wind while a negative correlation is made with the number of live load events with wind speeds. In addition, the target reliability index selected for wind plus vessel collision is lower than that used for live loads with winds.

3.3.7 Combination of Earthquakes and Scour (EQ + SC)

In this section, the single-column bent described in Figure 3.1 is analyzed to illustrate how the combined effects of earthquakes and scour will affect its reliability. The data from the five earthquake sites with probability distribution curves described in Figure 2.4 are used. The scour data are obtained from the USGS website for the Schohaire, Sandusky, and Rocky Rivers, as described in Section 2.4.4 (Table 2.8). The reliability calculations follow the Ferry-Borges model described in Section 2.5. The following assumptions are made:

- Scour depths last for 6 months, during which time the erosion of the soil around the column base remains at its highest value. The scour depth remains constant throughout the 6-month period.
- Scour depths are independent from year to year.
- Each site is on the average exposed to one major scour in 1 year.
- The probability distribution of the maximum yearly discharge rate follows a lognormal distribution. This maximum yearly discharge rate is assumed to control the scour depth for the year.
- The intensity of the earthquake response is as shown in Figure 2.15.
- The number of earthquakes expected in 1 year is 8 for the San Francisco site data, 2 for Seattle, 0.50 for Memphis, 0.40 for New York, and 0.01 for St. Paul. This means that there will be 600 earthquakes in a 75-year return period for San Francisco, 150 earthquakes for Seattle, 38 in Memphis, 30 in New York, and 1 in St. Paul.
- The probability distribution of the maximum yearly earthquake may be used to find the probability distribution for a single event using Equation 2.14.
- Equation 2.14 is used to find the probability distribution of the earthquake intensities for different return periods. In particular, the earthquake intensity for $t = 1/2$-year period is calculated for combination with the earthquake load effects when the foundation is scoured.
- The effects of each scour are combined with the effects of the $1/2$-year earthquake intensity and projected to provide the maximum expected combined load in the 75-year bridge design life period using the Ferry-Borges model described in Section 2.5. The calculations account for the cases in which the earthquake occurs with the $1/2$-year scour period and the cases in which an earthquake occurs when there is no scour.
- The reliability calculations are executed for the column overtipping limit state. The column bending limit state was not considered because the results of Figure 3.18 show that the foundation depth does not significantly affect the reliability index for column bending. Thus, when a large length of the column is exposed because of scour, the reduced column stiffness that ensues will result in lower bending moment on the column base, and the reliability index remains practically unchanged. On the other hand, for column overtipping, the remaining soil depth may sometimes not be sufficient to resist overtipping even though the forces are reduced.
- The reliability calculations account for the uncertainties associated with predicting the EQ intensity, estimating the bridge response given an EQ intensity, projecting the scour depth, and estimating the soil capacity to resist overtipping.
- The failure equation for bending of the column under the combined effect may be represented as follows:

94

$$Z = \max_{n_1} \left\{ \begin{array}{l} \max_{n_1} \left\{ \dfrac{3\gamma DK_p (L - y_{\text{scour}})^3}{6} - \max_{n_2} \left[F_{EQ}^* (e + L) \right] \right\} \\ \max_{n_3} \left\{ \dfrac{3\gamma DK_p (L)^3}{6} - F_{EQ}^{**} (e + L) \right\} \end{array} \right\} \quad (3.44)$$

where

γ = the specific weight of the soil;

D = the column diameter;

K_p = the Rankine coefficient;

L = the foundation depth before scour;

e = the distance from the original soil level to the point of application of the inertial force;

F_{EQ}^* = the inertial force when the scour is on;

F_{EQ}^{**} = the inertial force when no scour is on;

n_1 = the number of scours expected in a 75-year period (75 scours);

n_2 = the number of earthquakes expected in a $\frac{1}{2}$-year period when scour is on; and

n_3 = the number of earthquake events expected in a period equal to 75 years minus the times when a scour is on (i.e., n_3 is the number of earthquakes in a period equal to 37.5 years [75/2]).

The results of the reliability analysis are provided in Figure 3.34 for the San Francisco earthquake data for different values of foundation depth. The results that were obtained by considering the reliability of the bridge structure when only earthquakes are considered (i.e., by totally ignoring the effects

of scour) are also illustrated in Figure 3.34. The results for the San Francisco site are summarized as shown in Table 3.23. For example, if the bridge column is designed to satisfy the proposed revised AASHTO LRFD specifications (the seismic provisions) developed under NCHRP Project 12-49 (ATC and MCEER, 2002), then the required foundation depth for the column using a 2,500-year earthquake return period is $L = L_{\text{design}} = 26.06$ m (85.5 ft). This would produce a reliability index for overtipping (for a 75-year bridge design life) equal to $\beta = 2.26$ for sites having earthquake intensities similar to those observed in San Francisco but that are not subject to scour. If one considers that the foundation may be weakened by the presence of scour that would occur because of the river having the discharge rate of the Schohaire, then the reliability index for the column with a foundation depth of $L = 26.06$ m (85.5 ft) reduces to $\beta = 2.24$. If one wishes to increase the reliability index from 2.24 back to the original 2.26, then the depth of the foundation should be increased. By interpolation, the required foundation depth that will produce a reliability index equal to $\beta = 2.26$, when both earthquake loads and scour are considered, should be $L = 26.33$ m (86.4 ft) or an increase of 0.27 m (0.9 ft).

The L_{design} value 26.06 m (85.5 ft) includes the effect of the resistance factor for a lateral soil capacity of $\phi = 0.50$. The scour factor that should be used for the combination of earthquakes and scour should be determined such that

$$0.50 \frac{3\gamma DK_p L^\beta}{6} = \gamma_{EQ} F_{EQ} (e + \gamma_{SC} y_{SC} + L') \quad (3.45)$$

Reliability index for overtipping of one column under *EQ* and scour (San Francisco)

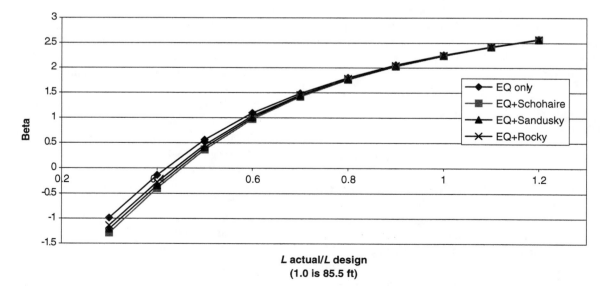

Figure 3.34. Reliability index for earthquakes and scour for overtipping (San Francisco).

TABLE 3.23 Summary of live load factors for combination of *EQ* plus *SC*

Earthquake site	Scour data	β target	Original foundation depth (ft)	Required foundation depth (ft)	Scour depth (ft)	γ_{sc}
	Schohaire			86.42	17.34	0.36
	Sandusky			86.19	14.33	0.36
San Francisco	Rocky	2.259	85.5	85.99	12.32	0.34
	Schohaire			67.04	17.34	0.18
	Sandusky			66.80	14.33	0.17
Seattle	Rocky	2.635	66	66.46	12.32	0.12
	Schohaire			28.64	17.34	0.55
	Sandusky			23.97	14.33	0.32
St. Paul	Rocky	2.368	20	22.08	12.32	0.20
	Schohaire			41.06	17.34	0.32
	Sandusky			39.39	14.33	0.23
New York	Rocky	2.491	37	38.55	12.32	0.18
	Schohaire			60.00	17.34	0.18
	Sandusky			59.59	14.33	0.16
Memphis	Rocky	2.488	58.6	59.20	12.32	0.12
Average						**0.25**

where

0.50 = the soil resistance factor;
γ = the specific weight of the soil;
D = the column diameter;
K_p = the Rankine coefficient;
L' = the foundation depth after scour ($L' = L_{required} - \gamma_{SC} y_{SC}$);
e = the distance from the original soil level to the point of application of the inertial force;
γ_{EQ} = the load factor for earthquake forces;
F_{EQ} = the inertial force;
y_{sc} = the design scour depth as shown in Table 3.4, which for the Schohaire River is 5.3 m (17.3 ft); and
γ_{SC} = the scour factor that should be used when considering the combination of earthquakes and scour.

It is noted that the inertial force, F_{EQ}, is a function of L' and the product $\gamma_{SC} y_{SC}$. Hence the scour factor can only be found from Equation 3.45 by trial and error.

The calibration of the load combination factor involves the determination of the values of γ_{SC} needed to produce the required foundation depth $L_{req} = 26.3$ m (86.4 ft). If the value for γ_{EQ} is preset at 1.0, then γ_{SC} is calculated to be equal to 0.36. This would lead to a foundation L' after scour erosion equal to 24.2 m (79.3 ft). Notice that the $L' = 24.2$ m (79.3 ft) is lower than the original $L = 26.1$ m (85.5 ft). This reduction in the required effective foundation depth is due to the additional flexibility of the column caused by the presence of scour. In fact, when scour occurs, the clear height of the column increases and the remaining foundation depth decreases; as long as the foundation is not totally washed out, this phenomenon would reduce the stiffness of the system and would lead to a higher natural period. When the natural period increases, the inertial forces decrease. Although the moment

arm of the force increases, the required foundation depth after scour would be lower than originally required if no scour were considered. It is noted however that adding 24.2 m (79.3 ft) plus 0.36 times the design scour depth (5.3 m = 17.3 ft) would still lead to a total foundation depth = 26.3 m (86.4 ft), which is higher than the original no scour depth $L = 26.1$ m (85.5 ft). It is also noted that in the examples solved in this section, the mass of the system is not altered because of the presence of scour or changes in foundation depth.

Similar calculations are executed for five earthquake sites and three discharge data. Namely, the earthquake data used correspond to those data from San Francisco, Seattle, Memphis, New York, and St. Paul. The scour data used correspond to those data from the Schohaire, Sandusky, and Rocky Rivers. The results from all 15 cases considered are summarized in Table 3.23. Notice that for all the cases considered, the scour factor remains below 0.55, with an average value of 0.25. Based on the results shown in Table 3.23, it would seem appropriate to recommend that a scour factor of $\gamma_{SC} = 0.25$ associated with an earthquake factor of $\gamma_{EQ} = 1.00$ be used to account for the combination of scour and earthquakes. The 0.25 scour factor is lower than that observed for the combination of scour and vessel collisions because the target reliability level used for the combination of scour and earthquakes is lower than that used for collisions and scour (e.g., for San Francisco, $\beta_{target} = 2.26$ versus $\beta_{target} = 2.78$ for vessel collisions). Another reason is because the earthquake analysis model accounts for the additional flexibility of the system when scour occurs and for the reduction in inertial force that this causes. The flexibility of the column is not directly considered when analyzing the effect of vessel collision forces.

All the analyses for combinations involving scour assume that scour lasts for a period of 6 months, during which time the scour remains at its highest value. Because no models are currently available to study the duration of live bed scours at bridge piers and how long it would take for the foundation of drilled shafts to regain their original strengths, a sensitivity analysis is performed to study the effect of scour duration on the results of the reliability analysis and the proposed load factor if the duration of scour is changed from 6 months to either 4 or 8 months. The results are provided in Table 3.24, which shows a change in the load factor on the order of 0.05 for each 2-month change in the scour duration. It is noted that in all cases considered, the scour is assumed to be at its highest value, which does provide a conservative envelope to the scour combination cases.

3.4 SUMMARY AND RECOMMENDATIONS

In this chapter, a reliability analysis of bridge bents is performed when the bridges are subjected to effects of live loads, wind loads, scour, earthquakes, vessel collisions, and combinations of these extreme events. Load combination factors are proposed such that bridges subjected to a combination of these events would provide reliability levels similar to those of bridges with the same configurations but situated in sites in which one threat is dominant. Thus, the proposed load factors are based on previous experiences with "safe bridge structures" and provide balanced levels of safety for each load combination. The results of this study found that different threats produced different reliability levels. Therefore, the target reliability indexes for the combination of events are selected in most cases to provide the same reliability level associated with the occurrence of the individual threat with the highest reliability index. Thus, when dealing with the combination of live load plus wind load or live load plus scour,

the reliability index associated with live loads is used as target. When studying the reliability of bridges subjected to the combination of wind loads and scour, the reliability index associated with wind loads alone is chosen for target. Similarly, when studying the reliability of bridges subjected to vessel collisions with scour or vessel collision with wind load, the reliability index associated with vessel collisions is used for target. For combinations involving earthquake loads, it is the reliability index associated with earthquakes alone that is used for target. Combinations involving earthquakes are treated differently than are other combinations because of the large additional capacity that would be required to increase the reliability levels of bridges subjected to earthquake risks. This approach is consistent with current earthquake engineering practice, which has determined that current earthquake design methods provide sufficient levels of safety given the costs that would be involved if higher safety levels were to be specified.

Results of the reliability analyses indicate that there are large discrepancies among the reliability levels implied in current design practices for the different extreme events under consideration. For example, the AASHTO LRFD was calibrated to satisfy a target member reliability index equal to 3.5 for gravity loads. The calculations performed in this study confirm that bridge column bents provide reliability index values close to the target 3.5 for the different limit states considered. These limit states include column bending, axial failure, bearing failure of the soil for one-column and multicolumn bents, and overtipping of one-column bents. The reliability index values calculated for each of these limit states show that β ranges from about 2.50 to 3.70. The lowest value is for foundation failures in bearing capacity. This is due to the large level of uncertainty associated with determining the strengths of foundation systems. The highest value is for bending

TABLE 3.24   Sensitivity of results to scour duration period

Earthquake site	Scour data	Required foundation depth (ft) (4 months)	Required foundation depth (ft) (6 months)	Required foundation depth (ft) (8 months)	γ_{sc} (4 months)	γ_{sc} (6 months)	γ_{sc} (8 months)
San Francisco	Schohaire	86.15	86.42	86.43	0.29	0.36	0.36
	Sandusky	86.01	86.19	86.38	0.30	0.36	0.42
	Rocky	85.87	85.99	86.07	0.29	0.34	0.38
Seattle	Schohaire	66.84	67.04	67.56	0.15	0.18	0.25
	Sandusky	66.68	66.80	67.24	0.15	0.17	0.25
	Rocky	66.16	66.46	66.75	0.04	0.12	0.19
St. Paul	Schohaire	28.58	28.64	28.68	0.55	0.55	0.56
	Sandusky	23.76	23.97	24.28	0.30	0.32	0.34
	Rocky	21.52	22.08	22.40	0.15	0.20	0.23
New York	Schohaire	40.15	41.06	41.52	0.25	0.32	0.35
	Sandusky	38.67	39.39	39.81	0.17	0.23	0.27
	Rocky	38.11	38.55	38.98	0.13	0.18	0.23
Memphis	Schohaire	59.85	60.00	60.57	0.16	0.18	0.25
	Sandusky	59.49	59.59	60.12	0.15	0.16	0.24
	Rocky	59.12	59.20	59.58	0.11	0.12	0.19
Average					*0.21*	*0.25*	*0.30*

moment capacity. These reliability indexes are for member failures. If there is sufficient redundancy, the system reliability is higher as explained by Ghosn and Moses (1998) and Liu et al. (2001).

The system reliability index for bridge bents subjected to earthquakes is found to be on the order of 2.80 to 3.00 for the moment capacity of drilled shafts supporting single- and multicolumn bents or 2.20 to 2.60 for overtipping of single-column bents. These values are for bridges designed following the proposed specifications developed under NCHRP Project 12-49 (ATC and MCEER, 2002]. These reliability levels decrease to as low as 1.75 when studying the safety of bridge columns under earthquakes. The large difference is due to the differences in the response modification factors recommended for use during the design process. In fact, *NCHRP Report 472* recommends the use of different response modification factors for bridge subsystems depending on the consequences of failure (ATC and MCEER, 2002). Unlike the analysis for other hazards, the earthquake analysis procedure accounts for system capacity rather than for member capacity because failure is defined by accounting for plastic redistribution of loads and the ductility capacity of the columns. Although relatively low compared with the member reliability index for gravity loads, the engineering community is generally satisfied with the current safety levels associated with current earthquake design procedures. For this reason, the target reliability index for load combination cases involving earthquakes is chosen to be the same reliability index calculated for the case in which earthquakes alone are applied. In fact, the results of this study indicate that very little improvement in the earthquake reliability index can be achieved even if large increases in the load factor are implemented. This is because much of the uncertainties in assessing the earthquake risk for bridge systems are due to the difficulty in predicting the earthquake intensities over the design life of the bridge. Improvement in the overall earthquake design process can only be achieved after major improvements in the seismologists' ability to predict future earthquakes.

The reliability index for designing bridge piers for scour for small-size rivers is on the order of 0.45 to 1.7, which is much lower than the 3.5 target for gravity loads and even lower than the range observed for earthquakes. In addition, failures caused by scour generally result in total collapse as compared with failures of members under gravity loads. This observation is consistent with the observation made by Shirole and Holt (1991) that, by far, most U.S. bridge collapses are due to scour. Because of the high risks of major collapses caused by scour, it is recommended to increase the reliability index for scour by applying a scour safety factor equal to 2.00. The application of the recommended 2.00 safety factor means that if current HEC-18 scour design procedures are followed, the final depth of the foundation should be 2.00 times the value calculated using the HEC-18 approach. Such a safety factor would increase the reliability index for scour from an average of about 1.00 to slightly more than 3.0, which

will make the scour design methods produce average safety levels for small rivers more compatible with the methods for other threats. However, the wide range in safety levels will require a review of current scour evaluation procedures.

Current AASHTO LRFD bridge design methods for wind loads provide average member reliability index values close to 3.00. However, there are large differences between the reliability indexes obtained for different U.S. sites. In fact, for the sites analyzed in this report, the reliability index (β) ranges between 2.40 and 4.00. The wind design approach is based on member safety. If sufficient levels of redundancy exist, the system reliability would be higher. The system reliability could increase by 0.25 to 0.50 over member reliability, depending on the number of columns in the bridge bent and the level of confinement of the columns. However, the large variations observed in the reliability index indicate that there should be major research effort placed on improving the current wind design procedures. This effort should be directed toward better understanding the behavior of bridges subjected to wind loads and toward developing new wind design maps that would provide more uniform safety levels for different regions of the United States.

The AASHTO vessel collision model produces a reliability index of about 3.15 for shearing failures and about 2.80 for bending failures. The higher reliability index for shear is due to the implicit biases and conservatism associated with the AASHTO LRFD shear design procedures as reported by Nowak (1999). The presence of system redundancy caused by the additional bending moment resistance by the bents, abutments, or both that are not impacted would increase the reliability index for bending failures to more than 3.00, making the safety levels more in line with those for shearing failures and those of bridge members subjected to the other threats considered.

The results of the reliability index calculations for individual threats are used to calibrate load combination factors applicable for the design of short- to medium-span bridges. The recommended load combination factors are summarized in Appendix A in a format that is implementable in the AASHTO LRFD specifications.

The load combination factors proposed in this study illustrate that the current load factors for the combination of wind plus gravity loads lead to lower reliability indexes than do those of either load taken separately. Hence, this study has recommended increasing the load factors for wind on structures and wind on live loads from the current 0.40 to 1.20 in combination with a live load factor of 1.00, which replaces the live load factor of 1.35. If the 1.35 live load factor is maintained, then the wind factor should be set equal to 0.80.

The commonly used live load factor equal to 0.50 in combination with earthquake effects leads to conservative results. This report has shown that a load factor of 0.25 on live load effects when they are combined with earthquake effects would still provide adequate safety levels for typical bridge configurations subjected to earthquake intensities similar to those observed on the West or the East Coast. These calculations are based on conservative assumptions on the recurrence of live loads when earthquakes are actively vibrating the bridge system.

For the combination of vessel collision forces and wind loads, a wind load factor equal to 0.30 is recommended in combination with a vessel collision factor of 1.0. The low wind load factor associated with vessel collisions compared with that recommended for the combination of wind loads plus live loads partially reflects the lower rate of collisions in the 75-year design life of bridges as compared with the number of live load events.

A scour factor equal to 1.80 is recommended for use in combination with a live load factor equal to 1.75. The lower scour load factor for combination of scour and live loads as compared with the load factor proposed for scour alone reflects the lower probability of having the maximum possible 75-year live load occur when the scour erosion is also at its maximum 75-year depth.

A scour factor equal to 0.70 is recommended in combination with a wind load factor equal to 1.40. The lower scour factor observed in combination involving wind loads as compared with those involving live loads reflect the lower number of wind storms expected in the 75-year design life of the structure.

A scour factor equal to 0.60 is recommended in combination with vessel collision forces. The lower scour factor observed in combinations that involve collisions reflects the lower number of collisions excepted in the 75-year bridge design life.

A scour factor equal to 0.25 is recommended in combination with earthquakes. The lower scour factor with earthquakes reflects the fact that as long as a total wash out of the foundation does not occur, bridge columns subjected to scour exhibit lower flexibilities that will help reduce the inertial forces caused by earthquakes. This reduction in inertial forces partially offsets the scour-induced reduction in soil depth and the resulting soil resisting capacity.

With regard to the extreme loads of interest to this study, the recommended revisions to the AASHTO LRFD specifications (1998) would address the extreme loads by ensuring that the factored member resistances are greater than the maximum load effects obtained from the following combinations:

- Strength I Limit State: $1.25\,DC + 1.75\,LL$
- Strength III Limit State: $1.25\,DC + 1.40\,WS$
- Strength V Limit State: $1.25\,DC + 1.00\,LL$
  $+\, 1.20\,WS + 1.20\,WL$
- Extreme Event I: $1.25\,DC + 0.25\,LL$
  $+\, 1.00\,EQ$
- Extreme Event II: $1.25\,DC + 0.25\,LL$
  $+\, 1.00\,CV$, or
  $1.25\,DC + 0.30\,WS$
  $+\, 1.00\,CV$ (3.46)

- Extreme Event III: $1.25\,DC;\ 2.00\,SC$,
  or
  $1.25\,DC + 1.75\,LL$;
  $1.80\,SC$
- Extreme Event IV: $1.25\,DC + 1.40\,WS$;
  $0.70\,SC$
- Extreme Event V: $1.25\,DC + 1.00\,CV$;
  $0.60\,SC$
- Extreme Event VI: $1.25\,DC + 1.00\,EQ$;
  $0.25\,SC$

In the above equations, DC represents the dead load effect, LL is the live load effect, WS is the wind load effect on the structure, WL is the wind load acting on the live load, EQ is the earthquake forces, CV is the vessel collision load, and SC represents the design scour depth. The dead load factor of 1.25 would be changed to 0.9 if the dead load counteracts the effects of the other loads.

Notice that no calculations for the combination of vessel collisions and live loads are performed in this study because no live load models are currently available to cover the long-span bridges most susceptible to this combination. The $\gamma_{LL} = 0.25$ factor proposed under Extreme Event II is projected from the calibration for the combination of earthquakes and live loads under Extreme Event I.

Unlike the other extreme events, scour does not produce a load effect. Scour changes the geometry of the system and reduces the load-carrying capacity of the foundation in such a way as to increase the risks from other failures. The presence of scour is represented in the above set of Equations 3.46 through the variable SC. The semicolon indicates that the analysis for load effects should assume that a maximum scour depth equal to $\gamma_{SC}\,SC$ exists when the load effects of the other extreme events are applied where g_{SC} is the scour factor by which scour depths calculated from the current HEC-18 method should be multiplied. When scour is possible, the bridge foundation should always be checked to ensure that the foundation depth exceeds $2.00\,SC$. For the cases involving a dynamic analysis, such as the analysis for earthquakes, it is very critical that the case of zero scour depth be checked because in many cases, the presence of scour may reduce the applied inertial forces. The resistance factors depend on the limit states being considered. When a linear elastic analysis of single and multicolumn bents is used, the system factors developed under NCHRP Project 12-47 should also be applied (Liu et al., 2001).

Equation 3.46 does not include any combinations of three different threats. This is because several analyses executed as part of this study have shown that the reductions of the reliability indexes for the combination of three different extreme events are small. This is due to the low probability of a simultaneous occurrence of three extreme events with high enough intensities to affect the overall risk. This is illustrated in Figure 3.33 for the combination of ship collisions with scour and winds loads and in Figure 3.35 for the combination of live load, wind, and scour.

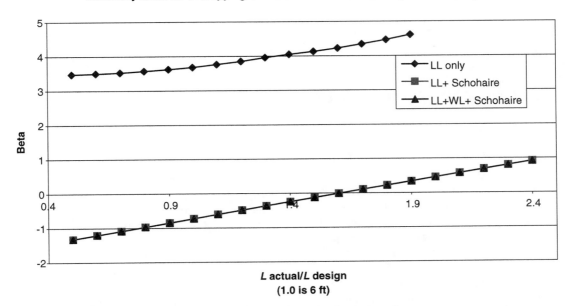

Figure 3.35. Reliability index for combination of live load, wind, and scour.

CHAPTER **4**

CONCLUSIONS AND FUTURE RESEARCH

4.1 CONCLUSIONS

This study has developed a design procedure for the application of extreme load events and the combination of their load effects in the *AASHTO LRFD Bridge Design Specifications* (1998). This is achieved by proposing a set of load factors calibrated using a reliability-based procedure that is consistent with the reliability methodology of the AASHTO LRFD specifications. The load events considered in this study include live loads, earthquakes, wind loads, ship collision loads, and scour. The reliability analysis of the effects of each load taken individually is performed using methods developed in previous bridge code calibration efforts (for the live loads and ship collisions) and during the development of other structural codes (for wind loads and earthquake loads). Because the current specifications for scour were not based on reliability methods, a scour reliability model has been developed for the purposes of this study. In addition, the Ferry-Borges model is used to evaluate the reliability of bridges under the combined effects of extreme load events. Results from the reliability of typical bridge configurations under the effects of individual threats are used to define target reliability levels for the development of load factors applicable for designing bridges that may be susceptible to a combination of threats. The objective is to recommend a rational and consistent set of load combination factors that can be implemented in future versions of the AASHTO LRFD specifications.

To achieve the objectives of the study, this project first reviewed the basic reliability methodology used during previous code calibration efforts. Basic bridge configurations designed to satisfy the current AASHTO specifications were analyzed to find the implicit reliability index values for different limit states for bridges subjected to live loads, wind loads, earthquakes, vessel collisions, or scour. The limit states considered include column bending, shearing failure, and axial failure of bridge columns, bearing failure of column foundations, and overtipping of single-column bents. The reliability analysis used appropriate statistical data on load occurrences and load intensities for the pertinent extreme events that were assembled from the literature and USGS websites. Statistical data on member and foundation capacities as well as load analysis models commonly used in reliability-based code calibration efforts were also used to find the probability of failure and the reliability index values for each extreme event.

Reliability indexes were calculated for the same bridges when subjected to the combination of extreme events using the Ferry-Borges model. The results were subsequently used to calibrate load combination factors appropriate for implementation in the LRFD equations.

The load factors are proposed such that bridges subjected to a combination of events provide reliability levels similar to those of bridges with the same configurations but situated at sites where one threat is dominant. Thus, the proposed load factors are based on previous experiences with "safe bridge structures" and provide balanced levels of safety for each load combination. Because this study found that different threats produced different reliability levels, the target reliability indexes for the combination of events are selected in most cases to provide the same reliability level associated with the occurrence of the individual threat with the highest reliability index. Thus, when dealing with the combination of live load plus wind load or live load plus scour, the reliability index associated with live loads is used as target. When studying the reliability of bridges subjected to the combination of wind loads and scour, the reliability index associated with wind loads alone is chosen for target. Similarly, when studying the reliability of bridges under vessel collision with scour or vessel collision with wind load, the reliability index associated with vessel collisions is used for target. For combinations involving earthquake loads, it is the reliability index associated with earthquakes alone that is used for target. Combinations involving earthquakes are treated differently than other combinations because of the large additional capacity that would be required to increase the reliability levels of bridges subjected earthquake risks.

The analysis considered structural safety as well as foundation safety. For two-column bents, system safety is compared with member safety. The results show that the system produces an additional reliability index about 0.25 higher than the reliability index of the individual members, which is consistent with the results of Liu et al. (2001) for drilled shafts of two-column bents formed by unconfined concrete columns. Hence, the system factors calibrated by Liu et al. (2001) are applicable for the cases in which linear elastic analysis is performed to check bridge member safety. Liu et al. (2001) calibrated system factors for application on the left-hand side of the design equation to complement the member resistance factor. The cases for which the application of system factors is possible include the analysis of bridges subjected to combinations exclusively involving live loads, wind loads, and ship collision forces. The analysis for combinations involving earthquakes is based on the plastic behavior of bridge bents; thus, system safety is directly considered and no system factors need to be applied. Scour causes the complete loss of the load-carrying capacity of a column, and bridge bents subjected to scour erosion exposing the full foundation will have little redundancy. Thus, such failures should be associated with system factors on the order of 0.80 as recommended by Liu et al. (2001).

Results of the reliability analyses indicate that there are large discrepancies among the reliability levels implied in current design practices for the different extreme events under consideration. Specifically, the following observations are made:

- The AASHTO LRFD was calibrated to satisfy a target member reliability index equal to 3.5 for gravity loads. The calculations performed herein confirm that members provide reliability index values close to the target 3.5 for the different limit states considered. These limit states include column bending, axial failure and overtipping of one-column bents. Lower reliability index values are observed for foundation-bearing capacities for one-column and multicolumn bents.
- The system reliability index for the drilled shaft foundations of bridge bents subjected to earthquakes is found to be on the order of 2.9 for moment capacity or 2.4 for overtipping of single-column bents. Even lower reliability values are observed for bridge columns because of the higher response modification factor recommended for column design as compared with those recommended for foundation subsystems. Unlike the analysis for other hazards, the earthquake analysis procedure accounts for system capacity rather than for member capacity. This is because earthquake analysis procedures consider the plastic redistribution of loads, and failure is defined based on the ductility capacity of the members. Although relatively low compared with the member reliability index for gravity loads, the engineering community is generally satisfied with the safety levels associated with current earthquake design procedures as improvements in the reliability index would entail high economic costs.
- The reliability index for designing bridge piers set in small rivers for scour varies from about 0.47 to 1.66, which is much lower than the 3.5 target for gravity loads. In addition, failures caused by scour result in total collapse as compared with failures of members under gravity loads that cause local damage only. Local damage can be sustained by the system if sufficient levels of redundancy and ductility are present, which is not the case for foundations exposed because of scour. Hence, it is recommended to increase the reliability index for scour by applying a scour safety factor equal to 2.00.

The application of the recommended 2.00 safety factor means that if current HEC-18 scour design procedures are followed, the final depth of the foundation should be 2.00 times the value calculated using the HEC-18 approach. Such a safety factor will increase the reliability index for scour from an average of about 1.0 to higher than 3.0. This increase will make the scour design methods more compatible with the methods for other threats.

- Although bridge design methods for wind loads provide an average member reliability index close to 3.0, there are large differences among the reliability indexes obtained for different U.S. sites. For this reason, it is recommended that research be undertaken to revise the existing wind maps so that they provide more consistent designs for different regions of the United States.
- The AASHTO vessel collision model produces a reliability index of about 3.15 for shearing failures and on the order of 2.80 for bending failures. The higher reliability index for shear is due to the higher implicit biases and conservative design methods. The presence of system redundancy caused by the reserve resistance provided by the bents, abutments, or both that are not impacted would increase the reliability index for bending failures to more than 3.00, making the safety levels more in line with those for shearing failures, which are brittle failures that do not benefit from the presence of redundancy.

The recommended load combination factors are summarized in Appendix A in a format that is implementable in the AASHTO LRFD specifications. The results illustrate the following points:

- The current load factors for the combination of wind plus live loads lead to lower reliability indexes than do those of either load taken separately. Hence, this study has recommended increasing the load factors for wind on structures and wind on live loads from the current 0.40 to 1.20 in combination with a live load factor of 1.00.
- The commonly used live load factor equal to 0.50 in combination with earthquake effects would lead to conservative results. This report has shown that a load factor of 0.25 on live load effects when they are combined with earthquake effects would still provide adequate safety levels for typical bridge configurations subjected to earthquake intensities similar to those experienced on the West and East Coasts. These calculations are based on conservative assumptions on the recurrence of live loads when earthquakes are actively vibrating the bridge system.
- For the combination of vessel collision forces and wind loads, a wind load factor equal to 0.30 is recommended in combination with a vessel collision factor of 1.0. The low wind load factor associated with vessel collisions compared with that recommended for the combination of wind loads plus live loads partially reflects the lower

rate of collisions in the 75-year design life of bridges as compared with the number of live load events.

- For the combination of vessel collisions and live load, it is recommended to reduce the live load factor from 0.50 to 0.25. This is proposed to bring this case more in line with the earthquake plus live load case. A higher wind load factor than live load factor is used in combinations involving vessel collisions to reflect the fact that the rate of vessel collisions increases during windstorms.

- A scour factor equal to 1.80 is recommended for use in combination with a live load factor equal to 1.75. The lower scour load factor for the combination of scour and live loads compared with the factor recommended for scour alone reflects the lower probability of having the maximum possible 75-year live load occur when the scour erosion is also at its maximum 75-year depth.

- A scour factor equal to 0.70 is recommended in combination with a wind load factor equal to 1.40. The lower scour factor observed for the combination with wind loads as compared with the combination with live loads reflects the lower number of wind storms expected in the 75-year design life of the structure.

- A scour factor equal to 0.60 is recommended in combination with vessel collision forces. The lower scour factor observed in combinations that involve collisions reflects the lower number of collisions expected in the 75-year bridge design life.

- A scour factor equal to 0.25 is recommended in combination with earthquakes. The lower scour factor with earthquakes reflects the fact that as long as a total wash out of the foundation does not occur, bridge columns subjected to scour exhibit lower flexibilities that will help reduce the inertial forces caused by earthquakes. This reduction in inertial forces partially offsets the scour-induced reduction in soil depth and the resulting soil resistance capacity.

- When scour is possible, the bridge foundation should always be checked to ensure that the foundation depth exceeds 2.0 times the scour depth obtained from the HEC-18 equations.

- For the cases involving a dynamic analysis such as the analysis for earthquakes, it is very critical that the case of zero scour depth be checked because in many cases, the presence of scour may reduce the applied inertial forces.

- When a linear elastic analysis of single-column and multicolumn bents is used, the system factors developed under NCHRP Project 12-47 should also be applied to complement the resistance factors (Liu et al., 2001).

4.2 FUTURE RESEARCH

The work performed as part of this study revealed that several issues related to the reliability analysis of bridge systems subjected to extreme load events need further investigation. These issues include the following areas discussed below.

4.2.1 Evaluation of Proposed Load Factors

This report has recommended several changes in the load factors for extreme events and their combinations. These recommendations are based on the reliability analysis of simplified models for typical bridge substructures subjected to individual extreme events and the combination of events. Extensive field evaluation and experimentation with the proposed recommendations using detailed structural analysis models should be undertaken. Some of these investigations may include extensive field measurements, data collection, and comparison of designs with evaluation of implied construction costs.

4.2.2 Determination of Appropriate Target Reliability Levels

The calibration of the load factors undertaken in this research follows a commonly used approach whereby the average reliability index from typical "safe" designs is used as the target reliability value for the new code. That is, a set of load factors and the nominal loads (or return periods for the design loads) are chosen for the new code such that bridges designed with these factors will provide reliability index values equal to the target value as closely as possible. This approach that has traditionally been used in the calibration of LRFD criteria (e.g., AISC and AASHTO) has led code writers to choose different target reliabilities for different types of structural elements or for different types of loading conditions. The choice of the different target βs raises the following question: If the reliability index (β) for live loads is 3.5 and for earthquake loads is 2.8, what should be the target when combining live loads and earthquake loads? In this report, it was decided that the higher reliability index should be used except for the cases involving earthquakes. The justification is that increasing the reliability index for earthquake threats would involve high construction costs, which society may not be willing to sustain. This subjective justification for using different target reliabilities can be formulated using a risk-benefit argument. For example, codes should tolerate a higher risk for the design of bridges (or structures) against a particular event if the costs associated with reducing this risk are prohibitive. Aktas, Moses, and Ghosn (2001) have presented examples illustrating an approach that can be used for determining the appropriate target reliability index values based on risk-benefit analysis. More work is needed in that direction to implement these concepts in actual bridge calibration efforts.

4.2.3 Reliability Models for Bridge Foundations

The analysis of bridge bents under the effect of the loads considered in this study is highly dependent on the accuracy of the foundation analysis models and the uncertainties associated with predicting the strength capacity of bridge foundations. Similarly, the effects of the soil-structure interaction on the response of bridge bents subjected to impact loading (i.e., vessel collisions) or cyclic dynamic loads (e.g., earthquakes and winds) need to be carefully considered. Currently, there is little information available to describe the uncertainty inherent in commonly used foundation analysis procedures. These uncertainties are caused by modeling assumptions, spatial variations of soil properties, and statistical uncertainties due to the limitations in the soil samples that are normally collected as part of the foundation design process. Some preliminary research work is currently ongoing under NCHRP Project 12-55. However, NCHRP Project 12-55 will not address all the aspects of the problem—particularly the dynamic effects of the loads on soil strength—and more research work is needed on this highly important subject. Reliability models should provide consistency between foundation analysis procedures and structural analysis models to better evaluate the reliability of complete structural systems and the interaction among bridge superstructures, substructures, and foundations.

4.2.4 Live Load Models

The statistical database used during the development of the live load model for the AASHTO LRFD code was collected from truck weight surveys in Ontario, Canada, and was supplemented by limited samples from Michigan. Some of the limitations in the data include the following:

1. The Ontario truck weight data were collected in 1975 in a Canadian province that had higher truck weight limits than the ones currently in effect in the United States.
2. The Ontario data were biased toward heavy trucks (i.e., only trucks that were believed to be heavily loaded were weighed); this would obviously produce a weight histogram showing an unusually high percentage of overloaded vehicles.
3. The Michigan truck weight data were also biased because Michigan allows higher truck weight limits for certain truck configurations than do most other U.S. states.
4. The statistics on multiple-truck occurrences used in the calibration were not based on specific observations and were unusually high compared with those observed on typical highways.
5. The average daily truck traffic used in the calibration was low compared with that observed on typical highways.
6. The same live load factor and live load model were proposed for all bridges, although it is widely accepted that bridges in rural areas with lower traffic counts are less likely to reach the projected maximum live load than are bridges in heavily traveled industrial regions.

In general, these assumptions had to be made during the original calibration because of lack of sufficient data. It is herein emphasized that, despite the limitations in the statistical database, the calibration process produces conservative estimates of the reliability levels and robust sets of load factors. However, the stated limitations in the database demonstrate the necessity of collecting more data in order to obtain a better assessment of the risks involved in current designs for various types of bridges and also to provide a mechanism to include site-specific information. The load combination factors depend on the number of live load events expected during the occurrence of the other loads. Hence, more information needs to be gathered on the rate and intensities of live load events. Future live load models should be sufficiently flexible to account for variations in these factors depending on the sites considered, including the legal truck weight limits in effect at the site and the intensity of the traffic.

4.2.5 Wind Load Models

Considerable concern has been expressed regarding the consistency of the data presented in the ASCE 7 maps that are the basis for the AASHTO maps for wind speed intensities. Some of the most pressing issues raised by researchers include the following:

1. Determining the proper probability distribution type that most adequately represents the intensity of wind speeds for hurricanes and for regular storms;
2. Considering statistical uncertainty and the effect of the tails of the probability distributions;
3. Examining the relationship between the actually measured wind speeds, including the regional variations of wind intensities, and the adequacy of the wind speed envelopes provided in the published maps;
4. Addressing the inconsistencies between the recurrence intervals used in the maps for hurricane winds and windstorms;
5. Remedying the lack of adequate models for representing special cases such as tornados; and
6. Developing models to account for wind gusts on bridges.

It is clear that the effect of winds on civil engineering structures in general and on bridge structures in particular did not receive the same attention given to other loads, and more work is needed in this field in order to develop rational and consistent design methods.

4.2.6 Scour Models

It is widely accepted that "a majority of bridges that have failed in the United States and elsewhere have failed due to scour" (AASHTO, 1994). This is confirmed by Shirole and Holt (1991), who observed that over a 30-year period, more than 1,000 of the 600,000 U.S. bridges have failed and that 60% of these failures are because of scour while earthquakes accounted for only 2%. Of course, there are many more bridges that are posted or otherwise taken out of service due to their inadequate strengths (because of deterioration, low rating, fatigue damage, etc.). Nevertheless, scour is considered to be a critical cause of failure because its occurrence often leads to total collapse. For these reasons, developing methods for the design and maintenance of bridge foundations for scour is currently considered to be a top priority for agencies concerned with the safety of bridges, and there is considerable research effort devoted to scour. The currently accepted model for scour design and evaluation is the HEC-18 model. HEC-18 stipulates that the scour depth produced by a given flood is not affected by previous (or existing) scour at the site. Hence, the maximum scour in a given return period is a function of the maximum flood observed in that period and is not affected by previous smaller floods that may have occurred within that same period. In addition, the HEC-18 model assumes that the flood duration is always long enough for the full scour depth corresponding to the flood velocity to be reached. Although, for live bed conditions, the scour hole is normally assumed to refill as the scour-causing flood recedes, the available literature does not provide precise information on how long it normally takes for the foundation to regain its original strength. This is believed to depend on the type of material being deposited by live-bed streams. It is also noted that HEC-18 was developed based on small-scale experiments involving sandy materials. Scale effects and the effects of different soil types need to be addressed in future research work. The importance of the scaling effect is reflected by the large differences observed in the reliability levels of rivers based on their discharge rates and expected scour depths. The differences between river scour and tidal scour should also be addressed. Simultaneously statistical data need to be gathered on each of the parameters that affect scour in order to develop a comprehensive reliability model for scour that is compatible with the models available for the other extreme events.

4.2.7 Vessel Collision Models

AASHTO's *Guide Specification and Commentary for Vessel Collision Design of Highway Bridges* developed a detailed, reliability-based model for studying the safety of bridges subjected to vessel collision forces (AASHTO, 1991). The model accounts for the major parameters that affect the rate of collisions and the magnitude of the collision forces. However, more data are needed to verify several of the assumptions used in the AASHTO model, including the effect of vessel size on the geometric probability of collision, and to correlate the rate of accidents with site-specific information on the type and size of the vessels, channel size and geometry, and other conditions. It is especially important to find explanations for the differences observed between the collision forces generated in laboratory experiments or computer analyses and actual damage observed after collisions in the field. These differences were lumped into a modeling random variable that was identified by the AASHTO guide specifications as the random variable $x = P_{actual}/P_{calculated}$. Efforts should be made to reduce the discrepancies between the predicted and observed forces to reasonable levels.

4.2.8 Earthquakes

Considerable effort has been expended over the last 3 decades on developing rational and consistent models for studying the safety of bridges subjected to earthquakes. To provide reasonable confidence levels, hazard maps and uniform hazard response spectra have been developed for a fine grid covering the whole United States. The issues that still require more research include the following:

1. Modeling of the ductility capacity and the relationship between ductility capacity and response modification factors, particularly for multidegree-of-freedom systems, taking into consideration the effects of the response modification factors on the overall reliability of bridge systems;
2. Development of SSI models that would provide consistent results for deep and shallow foundations;
3. Consideration of soil nonlinearity while determining the natural periods of the system;
4. Classification of soils for site amplification parameters and the consideration of uncertainties in determining the site factors.

4.2.9 Consideration of Modeling and Statistical Uncertainties

The research described in this report accounts for statistical and modeling uncertainties by representing these through random variables that are directly included in the calculations of the reliability index, β. For example, during the calibration of the load factors, different COV values are used to reflect the level of confidence associated with estimating the earthquake intensities at different sites. Similarly, different COV levels are used to reflect the number of data points used to estimate the mean and standard deviations of wind speeds at different sites throughout the United States. However, the final load factors proposed are averaged from all the sites and, thus, do not reflect the differences in the modeling and statistical uncertainties in a direct manner. Work on including uncertainty analysis in structural reliability formulation has been ongoing for a number of years (e.g., see the work of Ditlevsen [1982, 1988] and Der Kiureghian [2001]). However, there has not been a formal procedure that would explicitly account for the modeling and statistical uncertainties during the calibration of load factors and the development of structural design codes.

One possible approach would consist of calibrating two load factors for each load. One load factor would be a "generic load factor" that would be applicable for all sites and that would reflect the inherent randomness of the physical parameters describing the effects of the load under consideration. The second load factor would reflect the confidence level associated with the statistical data available to estimate the intensity of the loads at the particular site. The second load factor would also describe the difference between the results from the structural analysis and those observed in the field. Such an approach would encourage design engineers to collect more data on the loads and on the load intensities expected at the designated (or existing) bridge site and would also encourage the engineers to utilize more advanced analysis procedures or field measurements to reduce the modeling uncertainties associated with using simplified analysis and design methods. Moses (2001) has generally followed a similar approach during the calibration of the load and resistance factor rating and load capacity evaluation procedures for existing bridges. The possibility of employing the same format using a more formal analysis of uncertainty should be investigated.

REFERENCES

Aktas, E., Moses, F., and Ghosn, M. (2001). "Cost and Safety Optimization of Structural Design Specifications," *Journal of Reliability Engineering and System Safety*, Vol. 73, No. 3; pp. 205–212.

American Association of State Highway and Transportation Officials (1991). *Guide Specification and Commentary for Vessel Collision Design of Highway Bridges*, Washington, DC.

American Association of State Highway and Transportation Officials (1994). *AASHTO LRFD Bridge Design Specifications*, Washington, DC.

American Association of State Highway and Transportation Officials (1996). *Standard Specifications for Highway Bridges*, Washington, DC.

American Association of State Highway and Transportation Officials (1998). *AASHTO LRFD Bridge Design Specifications*, 2nd edition, Washington, DC.

American Concrete Institute (1995). *Building Code Requirements for Structural Concrete ACI 318-95*, Farmington Hills, MI.

American Institute of Steel Construction (1994). *Manual of Steel Construction: Load and Resistance Factor Design, LRFD*, 2nd edition, Chicago, IL.

American Society of Civil Engineers (1995). *Minimum Design Loads for Buildings and Other Structures*, Washington, DC.

Applied Technology Council and the Multidisciplinary Center for Earthquake Engineering Research (2002). *NCHRP Report 472: Comprehensive Specification for the Seismic Design of Bridges*, Transportation Research Board of the National Academies, Washington, DC.

Bea, R.G. (1983). "Characterization of the Reliability of Offshore Piles Subjected to Axial Loadings," in *Proceedings of the ASCE Structures Congress*.

Becker, D.E. (1996). "18th Canadian Geotechnical Colloqium: Limit States Design for Foundations. Part II Development for the National Building Code of Canada," *Canadian Geotechnical Journal*: Vol. 33; pp. 984–1007.

Belk, C.A, and Bennett, R.M. (1991). "Macro Wind Parameters for Load Combination," *ASCE Journal of Structural Engineering*, Vol. 117, No. 9.

Briaud, J.L., Ting, F.C.K., Chen, H.C., Gudavalli, R., Perugu, S., and Wei, G. (1999). "SRICOS: Prediction of Scour Rate in Cohesive Soils at Bridge Piers," *ASCE Journal of Geotechnical and Environmental Engineering*, Vol. 125, No. 4 (April 1999); pp. 237–246.

Chopra, A.K., and Goel, R.K. (2000). "Building Period Formulas for Estimating Seismic Displacements," *EERI Earthquake Spectra*, Vol. 16, No. 2.

Cowiconsult (1987). "General Principles for Risk Evaluation of Ship Collisions, Standings and Contact Incidents," unpublished technical report, January 1987.

Deng, L.Y., and Lin, D.K.J. (2000). "Random Number Generation for the New Century," *The American Statistician*, Vol. 54, No. 2; pp. 145–50.

Der Kiureghian, A. (2001). "Analysis of Structural Reliability under Model and Statistical Uncertainties: A Bayesian Approach," *Computational Structural Engineering*, Vol. 1, No. 2; pp. 81–87.

Ditlevsen, O. (1982). "Model Uncertainty in Structural Reliability," *Struc. Safety*, Vol. 1, No. 1; pp. 73–86.

Ditlevsen, O. (1988). "Uncertainty and Structural Reliability: Hocus Pocus or Objective Modeling?" Report No. 226, Department of Civil Engineering, Technical University of Denmark.

Ellingwood B., Galambos, T.V., MacGregor, J.G., and Comell, C.A. (1980). *Development of a Probability Based Load Criterion for American National Standard A58*, National Bureau of Standards, Washington, DC.

Frankel, A., Harmsen, S., Mueller, C., Barnhard, T., Leyendeker, E.V., Perkins, D., Hanson, S., Dickrnan, N., and Hopper, M. (1997). "USGS National Seismic Hazard Maps: Uniform Hazard Spectra, De-aggregation, and Uncertainty," *Proceedings of FHWA/ NCEER Workshop on the National Representation of Seismic Ground Motion for New and Existing Highway Facilities*, NCEER Technical Report 97-0010, SUNY Buffalo, NY; pp. 39–73.

Ghosn, M., and Moses, F. (1998). *NCHRP Report 406: Redundancy in Highway Bridge Superstructures*, Transportation Research Board of the National Academies, Washington, DC.

Goble, G.G., Burgess, C., Commander, B., Robson, B., and Schulz, J.X. (1991). "Load Prediction and Structural Response, Volume II"; Report to FHWA by Department of Civil and Architectural Engineering, University of Colorado, Boulder.

Haviland, R. (1976). "A Study of Uncertainties in the Fundamental Translational Periods and Damping Values for Real Buildings," MIT reports, Cambridge, MA.

Hwang, H.H.M., Ushiba, H., and Shinozuka, M. (1988). "Reliability Analysis of Code-Designed Structures under Natural Hazards," Report to MCEER, SUNY Buffalo, NY.

Hydraulic Engineering Center (1986). "Accuracy of Computed Water Surface Profiles," U.S. Army Corps of Engineers, Davis, CA.

International Association of Bridge & Structural Engineers (1983). "Ship Collision with Bridges and Offshore Structures," IABSE Colloquium, Copenhagen, 3 vols.

Johnson, P.A. (1995). "Comparison of Pier Scour Equations Using Field Data," *ASCE Journal of Hydraulic Engineering*, Vol. 121, No. 8; pp. 626–629.

Johnson, P.A., and Dock, D.A., (1998). "Probabilistic Bridge Scour Estimates," *ASCE Journal of Hydraulic Engineering*, Vol. 124, No. 7; pp. 750–755.

Landers, M.N., and Mueller, D.S. (1996). "Channel Scour at Bridges in the United States," FHWA-RD-95-l84, Federal Highway Administration, Turner-Fairbank Highway Research Center, McLean, VA.

Larsen, O.D. (1993). "Ship Collision with Bridges," Structural engineering documents, International Association for Bridge and Structural Engineering, Zurich.

Liu, D., Neuenhoffer, A., Chen, X., and Imbsen, R. (1998). "Draft Report on Derivation of Inelastic Design Spectrum," Report to NCEER, SUNY Buffalo, NY.

Liu, W.D., Neuenhoffer, A., Ghosn, M., and Moses, F. (2001). *NCHRP Report 458: Redundancy in Highway Bridge Substructures*, Transportation Research Board of the National Academies, Washington, DC.

Melchers, R.E. (1999). *Structural Reliability: Analysis and Prediction*, John Wiley & Sons, New York, NY.

Miranda, E. (1997). "Strength Reduction Factors in Performance-Based Design," EERC-CUREe Symposium in Honor of Vitelmo V. Bertero, Berkeley, CA.

Moses, F., and Ghosn, M. (1985). "A Comprehensive Study of Bridge Loads and Reliability," Final Report to Ohio DOT, Columbus, OH.

Moses, F. (1989). "Effects on Bridges of Alternative Truck Configurations and Weights," unpublished report for a truck weight study to the Transportation Research Board.

Moses, F. (2001). *NCHRP Report 454: Calibration of Load Factors for LRFR Bridge Evaluation,* Transportation Research Board of the National Academies, Washington, DC.

National Earthquake Hazards Reduction Program (1997). *Recommended Provisions for Seismic Regulations for New Buildings and Other Structures,* Federal Emergency Management Agency, FEMA 302, Building Safety Council, Washington, DC.

Newmark, N.M., and Hall, W.J. (1973). "Seismic Design Criteria for Nuclear Reactor Facilities," *Building Practices for Disaster Mitigation, Report No. 46,* National Bureau of Standards, U.S. Department of Commerce; pp. 209–236.

Nowak, A.S., and Hong, Y.K. (1991). "Bridge Live Load Models," *ASCE Journal of Structural Engineering,* Vol. 117, No. 9.

Nowak, A.S. (1999). *NCHRP Report 368: Calibration of LRFD Bridge Design Code,* Transportation Research Board of the National Academies, Washington, DC.

Nowak, A.S., and Collins, K.R. (2000). *Reliability of Structures,* McGraw Hill, New York, NY.

Poulos, H.G., and Davis, E.H. (1980). *Pile Foundation Analysis and Design,* Krieger Publishing Co., FL.

Priestley, M.J.N., and Park, R. (1987). "Strength and Ductility of Concrete Bridge Columns Under Seismic Loading," *ACI Structural Engineering Journal,* January–February.

Richardson, E.V., and Davis, S.R., (1995). *Evaluating Scour at Bridges,* 3rd edition. Report No. FHWA-IP-90-017, Hydraulic Engineering Circular No. 18, Federal Highway Administration, Washington, DC.

Seed, H.B., Ugas, C., and Lysmer, J. (1976). "Site-Dependant Spectra for Earthquake-Resistant Design," *Bulletin of the Seismological society of America,* Vol. 66, No. 1; pp. 221–243.

Shirole, A.M., and Holt, R.C. (1991). "Planning for a Comprehensive Bridge Safety Assurance Program," *Transportation Research Record 1290,* Transportation Research Board of the National Academies, Washington, DC; pp. 39–50.

Simiu, E., Changery, M.J., and Filliben, J.J. (1979). *Extreme Wind Speeds at 129 Stations in the Contiguous United States,* National Bureau of Standards, Washington, DC.

Simiu, E., and Heckert, N.A. (1995). *Extreme Wind Distribution Tails: A Peaks over Threshold Approach,* NIST Building Science Series, No. 174, Washington, DC.

Stewart, J.P., Seed, R.B., and Fenves, G.L. (1999). "Seismic Soil Structure Interaction in Buildings, II: Empirical Findings." *ASCE Journal of Geotechnical and Geoenvironmental Engineering,* Vol. 125, No. 1.

Takada, T., Ghosn, M., and Shinozuka, M. (1989). "Response Modification Factors for Buildings and Bridges," *Proceedings from the 5th International Conference on Structural Safety and Reliability,* ICOSSAR 1989, San Francisco; pp. 415–422.

Thoft-Christensen, P., and Baker, M. (1982). *Structural Reliability Theory and its Applications,* Springer Verlag, Berlin, Germany.

Turkstra, C.J., and Madsen, H. (1980). "Load Combinations in Codified Structural Design," *ASCE, Journal of Structural Engineering,,* Vol. 106, No. 12; pp. 2527–2543.

U.S. Nuclear Regulatory Commission (1989). *Standard Review Plan, NUREG-O800,* Rev. 2, U.S. Department of Energy, Washington, DC.

Wen, Y.K. (1977). "Statistical Combination of Extreme Loads," *ASCE Journal of Structural Engineering,* Vol. 103, No. 5; pp. 1079–1095.

Wen, Y.K. (1981). "A Clustering Model for Correlated Load Processes," *ASCE Journal of Structural Engineering,* Vol. 107, No. 5; pp. 965–983.

Whalen, T.M., and Simiu, E. (1998). "Assessment of Wind Load Factors for Hurricane-Prone Regions," *Journal of Structural Safety,* Vol. 20; pp. 271–281.

Whitney, M.W., Barik, I.E., Griffin J.J., and Allen, D.L. (1996). "Barge Collision Design of Highway Bridges," *ASCE Journal of Bridge Engineering,* Vol. I, No. 2; pp. 47–58.

Zahn, F.A., Park, R., and Priestly, M.J.N. (1986). "Design of Reinforced Concrete Bridge Columns for Strength and Ductility," Report 86- 7, University of Canterbury, Christchurch, New Zealand.

GLOSSARY OF NOTATIONS

A = Earthquake peak ground acceleration.

A_0 = Cross-sectional area of the stream.

a_B = Barge bow damage depth.

AF = Annual failure rate.

b = River channel width.

BR = Vehicular braking force.

BR_a = Aberrancy base rate.

b_x = Bias of $\bar{x} = x/x_n$.

c = Wind analysis constant.

C' = Response spectrum modeling parameter.

CE = Vehicular centrifugal force.

C_F = Cost of failure.

C_H = Hydrodynamic coefficient that accounts for the effect of surrounding water on vessel collision forces.

C_I = Initial cost for building bridge structure.

C_p = Wind pressure coefficient.

CR = Creep.

C_T = Expected total cost of building bridge structure.

CT = Vehicular collision force.

CV = Vessel collision force.

D = Diameter of pile or column.

DC = Dead load of structural components and nonstructural attachments.

DD = Downdrag.

D_n = K–S maximum difference between measured cumulative probability and expected probability.

DW = Dead load of wearing surfaces and utilities.

e = Height of column above ground level.

EH = Horizontal earth pressure load earth surcharge load.

E_p = Modulus of elasticity for pile.

EQ = Earthquake.

EV = Vertical pressure from dead load of earth fill.

E_z = Wind exposure coefficient.

f = Location of maximum bending moment in pile shaft below the soil surface.

F_0 = Froude number.

F_a = Site soil coefficients for short periods.

F_{apl} = Equivalent applied force.

F_i = Equivalent inertial force.

F_K = Applied force for load type K.

FR = Friction.

F_v = Site soil coefficient for system with natural period $T = 1$ sec.

$F_Y(Y^*)$ = Cumulative probability = the probability that the variable Y takes a value less than or equal to Y^*.

G = Wind gust factor.

g = Acceleration caused by gravity.

H = Moment arm of applied force.

HL-93 = AASHTO LRFD design live load.

IC = Ice load.

I_{IM} = Dynamic amplification for live load.

I_{LL} = Live load intensity in terms of HL-93.

IM = Vehicular dynamic load allowance.

I_p = Moment of inertia of pile.

$I_{\rho H}$ = Soil influence coefficient for lateral force.

$I_{\rho M}$ = Soil influence coefficient for moment.

$K_1, K_2,$
$K_3,$ and K_4 = Scour coefficients that account for the nose shape of the pier, the angle between the direction of the flow and the direction of the pier, the streambed conditions, and the bed material size.

K_p = Rankine coefficient.

K_R = Pile flexibility factor, which gives the relative stiffness of the pile and soil.

L = Foundation depth.

L_e = Effective depth of foundation (distance from ground level to point of fixity).

LL = Vehicular live load.

LOA = Overall length of vessel.

LS = Live load surcharge.

$\max(x)$ = Maximum of all possible x values.

M_{cap} = Moment capacity.

M_{col} = Moment capacity of column.

M_{design} = Design moment.

n = Manning roughness coefficient.

N_i = Number of vessels (or flotillas) of type i.

PA = Probability of aberrancy.

P_B = Nominal design force for ship collisions.

P_{B0} = Base wind pressure.

P_{cap} = Axial force capacity.

$PC_{i,k}$ = Probability that bridge will collapse given that a vessel of type i has collided with bridge member k.

P_{col} = Axial capacity of column.

P_{design} = Design axial force.

P_f = Probability of failure.

PG = Geometric probability.

PL = Pedestrian live load.

P_p = Passive resultant resisting force of the soil.

P_{soil} = Soil bearing capacity.

Q = River flow discharge rate.

R = Resistance or member capacity.

R_B = Correction factor for impacting barge width.

R_{Ba} = Correction factor for bridge location for vessel aberrancy.

R_c = Correction factor for current acting parallel to vessel path.

R_D = Correction factor for vessel traffic density.

R_H = Hydraulic radius.

R_m = Response modification factor.

R_{xc} = Correction factor for crosscurrents acting perpendicular to vessel path.

S = Load effect.

S_0 = Slope of the river bed stream.

S_a = Spectral acceleration.

SC = Scour.

S_{D1} = Spectral acceleration for a natural period $T_1 = 1$ sec.

S_{Ds} = Spectral acceleration for short period $T_s = 0.2$ sec.

SE	=	Settlement.
SH	=	Shrinkage.
T	=	Natural period of the system.
t'	=	Natural period modeling factor.
T_0	=	Natural period at which the maximum spectral acceleration is reached.
TG	=	Temperature gradient.
T_s	=	Natural period at which the spectral acceleration begins to decrease.
TU	=	Uniform temperature.
V	=	Velocity (for wind speed, vessels at impact, or river flow).
V_0	=	Wind friction velocity.
V_{10}	=	Wind velocity above ground level.
V_B	=	Base wind velocity = 160 km/h (100 mph).
V_{col}	=	Shear capacity of column.
V_{DZ}	=	Design wind velocity at design elevation Z.
V_x	=	Coefficient of variation (COV) of *x* = standard deviation/mean value.
W	=	Weight (for vessel or structure).
w	=	Vessel weight modeling variable.
WA	=	Water load and stream pressure.
WL	=	Wind on live load.
WS	=	Wind load on structure.
x	=	Vessel collision modeling variable.
$\bar{x}$	=	Mean value of random variable *x*.
$X_{max,T}$	=	Maximum value of variable *X* in a period of time *T*.
x_n	=	Nominal value of *x* as specified by design code.

y_0	=	Depth of river flow just upstream of bridge pier excluding local scour.
y_{max}	=	Maximum depth of scour.
y_{scour}	=	Scour depth.
Z	=	Safety margin = R–S.
Z_0	=	Friction length for wind.
β_{target}	=	Target reliability index used for calibration.
β	=	Reliability index.
ε	=	Standard error in regression equation.
ϕ	=	Resistance factor.
ϕ_s	=	Angle of friction for sand.
Φ	=	Cumulative standard normal distribution function.
Φ_0	=	Unit adjustment parameter = 1.486 for U.S. units or = 1.0 for SI units.
γ	=	Specific weight of sand.
γ_k	=	Load factor for load type *K*.
λ_{cyc}	=	Variable representing cyclic effects.
λ_{eq}	=	Modeling factor for the analysis of earthquake loads.
λ_{LL}	=	Live load modeling factor.
λ_{sc}	=	Scour modeling variable.
λ_{sys}	=	System factor that represents the capacity of the "system" to continue to carry loads after failure of first member.
λ_V	=	Statistical modeling for estimating wind speed *V*.
μ_{cap}	=	Ductility capacity of a concrete column.
$\mu_{specified}$	=	Specified ductility capacity.
ν_i	=	Yearly rate of collisions for each vessel (or flotilla) of type *i*.
σ_x	=	Standard deviation of a random variable *x*.

APPENDIXES

INTRODUCTION

Appendixes A, B, C, H, and I are published herein and on *CRP-CD-30;* Appendixes D through G are published only on *CRP-CD-30.* Appendixes D through G have not been edited by TRB.

Appendix A gives the recommended revisions to the AASHTO LRFD specifications. Proposed modifications to text are underlined. Proposed modifications to tables are shaded.

Appendixes B through G describe and illustrate the reliability and analysis models used during the calibration of the proposed load combination factors. Appendixes B through G were written by Michel Ghosn from the City University of New York; Peggy Johnson from Penn State University; Fred Moses from the University of Pittsburgh; Jian Wang, Research Assistant at the City University of New York; David Liu from Imbsen & Associates, Inc.; Darrel Gagnon from Buckland & Taylor; and Mark Hunter from Rowan, Williams, Davies, & Irwin, Inc.

Appendixes H and I describe and illustrate the earthquake and scour reliability models developed during a scope extension phase of NCHRP Project 12-48. Appendix I performs a reliability analysis of a bridge that has the same configuration of an example bridge analyzed as part of NCHRP Project 12-49, which developed a set of specifications for the seismic design of highway bridges and was published as *NCHRP Report 472.* The objective of this appendix is to verify that the results presented in other appendixes of *NCHRP Report 489* that use simplified analysis procedures are consistent with those obtained with more complete analysis methods. Appendix I develops a method to perform the reliability analysis

of scour based on the data collected by Landers and Mueller in their report entitled "Channel Scour at Bridges in the United States" (FHWA-RD-95-184). Appendixes H and I were written by Michel Ghosn from the City University of New York; Fred Moses, Engineering Consultant; and Jian Wang, Research Assistant at the City University of New York. The contributions of Charles Annis from Statistical Engineering, Inc., whose help and advice were instrumental for completing the scour model, are gratefully acknowledged.

The appendixes and their authorship are as follows.

- Appendix A: Recommended Modifications to AASHTO LRFD Bridge Design Specifications
- Appendix B: Reliability Model for Scour Analysis (Peggy Johnson)
- Appendix C: Reliability Analysis of Three-Span Bridge Model (Michel Ghosn, Fred Moses, and Jian Wang)
- Appendix D: Reliability Analysis of I-40 Bridge Under the Combined Effect of Scour and Earthquakes (Michel Ghosn, David Liu, and Peggy Johnson)
- Appendix E: Analysis of Maysville Bridge for Vessel Collision (Michel Ghosn and Fred Moses)
- Appendix F: Analysis of Maysville Bridge for Earthquakes (Darrel Gagnon)
- Appendix G: Analysis of Maysville and I-40 Bridges for Wind Loads (Mark Hunter)
- Appendix H: Seismic Risk Analysis of a Multispan Bridge (Michel Ghosn, Fred Moses, and Jian Wang)
- Appendix I: Analysis of Scour Data and Modified Reliability Model for Scour (Michel Ghosn, Fred Moses, and Jian Wang)

APPENDIX A

RECOMMENDED MODIFICATIONS TO AASHTO LRFD BRIDGE DESIGN SPECIFICATIONS

Section 3 -loads and load Factors (SI)

SPECIFICATIONS

COMMENTARY

3.3.2 Load and Load Designation

The following permanent and transient loads, forces, and effects of extreme events shall be considered:

• Permanent Loads

DD = downdrag
DC = dead load of structural components and nonstructural attachments
DW = dead load of wearing surfaces and utilities
EH = horizontal earth pressure load
ES = earth surcharge load
EV= vertical pressure from dead load of earth fill

• Transient Loads

BR = vehicular braking force
CE = vehicular centrifugal force
CR = creep
CT = vehicular collision force
CV = vessel collision force
EQ = earthquake
FR = friction
IC = ice load
IM = vehicular dynamic load allowance
LL = vehicular live load
LS = live load surcharge
PL = pedestrian live load
SC = scour
SE = settlement
SH = shrinkage
TG = temperature gradient
TU = uniform temperature
WA = water load and stream pressure
WL = wind on live load
WS = wind load on structure

Scour is not a load but is an extreme event that alters the geometry of the structure and foundation possibly causing structural collapse or the amplification of the effects of applied loads.

3.4 LOAD FACTORS AND COMBINATIONS

3.4.1 Load Factors and Load Combinations

The total factored load shall be taken as:

$$Q = \eta \sum \gamma_i q_i \qquad (3.4.1-1)$$

where:

η = load modifier specified in Article 1.3.2
q_i = loads specified herein
γ_i = load factors specified in Tables 1 and 2

C3.4.1

The background for the load factors specified herein, and the resistance factors specified in other sections of these Specifications are developed in Nowak (1999) and Ghosn, Moses, and Wang (2003).

An alternative formulation that replaces the load modifier by a system factor placed on the left-hand side of the design equation has been proposed by Ghosn and Moses (1998) and Liu, Ghosn, Moses and Neuenhoffer, A. (2001).

Section 3 -Loads and Load Factors (SI)

SPECIFICATIONS

Components and connections of a bridge shall satisfy Equation 1.3.2.1-1 for the applicable combinations of factored extreme force effects as specified at each of the following limit states:

- STRENGTH I-

Basic load relating to vehicular use without wind.

- STRENGTH II -

Load combination relating to the use of the bridge by Owner specified special design vehicles and/or evaluation permit vehicles, without wind.

- STRENGTH III -

Load combination relating to the bridge exposed to wind velocity exceeding 90 km/hr.

- STRENGTH IV -

Load combination relating to very high dead load to live load force effect ratios.

COMMENTARY

A reduced value of 0.50, applicable to all strength load combinations, specified for TU, CR and SH, used when calculating force effects other than displacements at the strength limit state, represents an expected reduction of these force effects in conjunction with the inelastic response of the structure. The calculation of displacements for these loads utilizes a factor greater than 1.0 to avoid undersized joints and bearings. The effect and significance of the temperature gradient remains unclear at this writing. Consult Article C3.12.3 for further information.

The permit vehicle should not be assumed to be the only vehicle on the bridge unless so assured by traffic control. Otherwise, the other lanes should be assumed to be occupied by the vehicular live load as specified herein. For bridges longer than the permit vehicle, the presence of the design lane load, preceding and following the permit load in its lane, should be considered.

Vehicles become unstable at higher wind velocities. Therefore, high winds prevent the presence of significant live load on the bridge.

The standard calibration process for the strength limit state consists of trying out various combinations of load and resistance factors on a number of bridges and their components. Combinations which yield a safety index close to the target value of $\beta = 3.5$ are retained for potential application. From these are selected constant load factors γ and corresponding resistance factors β for each type of structural component reflecting its use.

This calibration process had been carried out for a large number of bridges with spans not exceeding 60 000 mm. For the primary components of large bridges, the ratio of dead and live load force effects is rather high, and could result in a set of resistance factors different from those found acceptable for small- and medium-span bridges. It is believed to be more practical to investigate one more load case than to require the use of two sets of resistance factors with the load factors provided in Strength Load Combination I, depending on other permanent loads present. Spot checks had been made on a few bridges with up to 183 000 mm spans, and it appears that Load Combination IV will govern where the dead load to live load force effect ratio exceeds about 7.0.

Section 3 -Loads and Load Factors (SI)

SPECIFICATIONS

- **STRENGTH V -**

Load combination relating to normal vehicular use of the bridge with wind of 90 km/hr velocity.

COMMENTARY

- **EXTREME EVENT I –**

Load combination including earthquake.

This limit state includes live loads, LL in combination with water pressure and earthquakes. The probability of a major earthquake occurring during the crossing of maximum vehicular live loads is small. Hence the use of the 0.25 live load factor is justified.

- **EXTREME EVENT II -**

Load combination relating to ice load, collision by vessels and vehicles, and to certain hydraulic events with reduced live load, other than that which is part of the vehicular collision load, CT.

The joint probability of these events is extremely low, and, therefore, they are specified to be applied separately. Under these extreme conditions, the structure is expected to undergo considerable inelastic deformation by which locked-in force effects due to TU, TG, CR, SH and SE are expected to be relieved.

The 0.25 live load factor reflects a low probability of the concurrence of the maximum vehicular live load, and the extreme events.

Similarly, a 0.30 wind load factor reflects a low probability of the occurrence of maximum wind speeds and the extreme events.

Because of the very low probability of simultaneous occurrence of large live loads, high winds, and the extreme events, the designer does not need to account for the simultaneous combination of all three effects.

- **EXTREME EVENT III**

Load combination relating to live load and scour.

A check of the safety of the foundation should first be performed by multiplying the scour depth calculated from the HEC-18 equations by a scour safety factor equal to 2.00. This 2.00 factor ensures that bridges susceptible to scour will still have reliability levels similar to those of other bridges.

This limit state includes live loads, LL, water loads, WA, in the presence of scour, SC. The probability that the maximum scour depth is present at the same time that maximum live loads are on the bridge is significant. Therefore, 1.80 times the maximum calculated scour depth should be used in combination with the maximum factored live load.

Section 3 -Loads and Load Factors (SI)

SPECIFICATIONS

COMMENTARY

- EXTREME EVENT IV

Load combination relating to wind loads in the presence of scour.

This limit state includes water loads, WA, and wind loads on structures, WS, in the presence of scour, SC. To account for the probability of a major windstorm occurring in the presence of the maximum scour , 70% of the calculated scour depth should be used in combination with wind loads.

- EXTREME EVENT V

Load combination relating to ice load, collision by vessels or vehicles in the presence of scour.

This limit state includes water loads, WA, in the presence of scour, SC, in combination with either ice loads, collisions by vessels, or vehicles. The probability of a collision or maximum ice loads occurring in the presence the maximum scour depth is small. Therefore, only 60% of the calculated scour depth should be used in combination with either ice load or a collision.

- EXTREME EVENT VI

Load combination relating to earthquakes with scour.

This limit state includes water loads, WA and earthquakes in the presence of scour, SC. The probability of an earthquake occurring in the presence of the maximum scour depth is small. Therefore, consideration of using only 25% of design scour depth may be warranted. Similarly, consideration of basing water loads on mean discharges may be warranted.

The presence of scour may increase the natural period of the bridge system and lead to lower inertial forces. Hence, the worst case scenario from Extreme Events I and VI should be utilized.

The calibration of the scour combination factors for Extreme Events III through VI has been performed by Ghosn, Moses and Wang (2003) for bridge foundations set in river channels assuming independence between scour and other extreme events.

116

Section 3 -Loads and Load Factors (SI)

SPECIFICATIONS

- SERVICE I -

Load combination relating to the normal operational use of the bridge with 90 km/hr wind, and with all loads taken at their nominal values. Also related to deflection control in buried metal structures, tunnel liner plate and thermoplastic pipe and to control crack width in reinforced concrete structures.

- SERVICE II -

Load combination intended to control yielding of steel structures and slip of slip- critical connections due to vehicular live load.

- SERVICE III -

Load combination relating only to tension in prestressed concrete with the objective of crack control.

COMMENTARY

Compression in prestressed concrete components is investigated using this load combination. Service III is used to investigate tensile stresses in prestressed concrete components.

This load combination corresponds to the overload provision for steel structures in past editions of the AASHTO Specifications, and it is applicable only to steel structures. From the point of view of load level, this combination is approximately halfway between that used for Service I and Strength I Limit States.

The live load specified in these Specifications reflects, among other things, current exclusion weight limits mandated by various jurisdictions. Vehicles permitted under these limits have been in service for many years prior to 1993. There is no nationwide physical evidence that these vehicles have caused detrimental cracking in existing prestressed concrete components. The statistical significance of the 0.80 factor on live load is that the vent is expected to occur about once a year for bridges with two traffic lanes, less often for bridges with more than two traffic lanes, and about once a day for the bridges with a single traffic lane.

Section 3 -Loads and Load Factors (SI)

SPECIFICATIONS

- FATIGUE -

Fatigue and fracture load combination relating to repetitive gravitational vehicular live load and dynamic responses under a single design truck having the axle spacing specified in Article 3.6.1.4.1.

The load factors for various loads comprising a design load combination shall be taken as specified in Table 1. All relevant subsets of the load combinations shall be investigated. For each load combination, every load that is indicated to be taken into account and which is germane to the component being designed, including all significant effects due to distortion, shall be multiplied by the appropriate load factor and multiple presence factor specified in Article 3.6.1.1.2 if applicable. The products shall be summed as specified in Equation 1.3.2.1-1 and multiplied by the load modifiers specified in Article 1.3.2.

The factors shall be selected to produce the total extreme factored force effect. For each load combination, both positive and negative extremes shall be investigated.

In load combinations where one force effect decreases the effect of another, the minimum value shall be applied to the load reducing the force effect. For permanent force effects, the load factor which produces the more critical combination shall be selected from Table 2. Where the permanent load increases the stability or load-carrying capacity of a component or bridge, the minimum value of the load factor for that permanent load shall also be investigated.

The larger of the two values provided for load factors of TU, CR and SH shall be used for deformations and the smaller values for all other effects.

For the evaluation of overall stability of earth slopes with or without a foundation unit, only the maximum load factors shall be used.

COMMENTARY

The load factor, applied to a single design truck, reflects a load level which has been found to be representative of the truck population with respect to a large number of return cycles of stresses, and their cumulative effects in steel elements, components and connections.

This article reinforces the traditional method of selecting load combinations to obtain realistic extreme effects, and is intended to clarify the issue of the variability of permanent loads and their effects. As has always been the case, the Owner or Designer may determine that not all of the loads in a given load combination apply to the situation under investigation.

It is recognized herein that the actual magnitude of permanent loads may also be less than the nominal value. This becomes important where the permanent load reduces the effects of the transient ones.

It has been observed that the probability of permanent loads exceeding the nominal value is larger than the probability of being less.

In the application of permanent loads, force effects for each of the specified six load types should be computed separately. Assuming variation of one type of load by span, length or component within a bridge is not necessary. For example, when investigating uplift at a bearing in a continuous beam, it would not be appropriate to use the maximum load factor for permanent loads in spans which produce a negative reaction and the minimum load factor in the spans which produce a positive reaction. Consider the investigation of uplift. Uplift was treated as a separate load case in past editions of the AASHTO Standard Specifications, but now becomes a strength load combination. Where a permanent load produces uplift, that load would be multiplied by the maximum load factor, regardless of the span in which it is located. If another permanent load reduces the uplift, it would be multiplied by the minimum load factor, regardless of the span in which it is located. For example, at Strength I Limit State where the permanent load reaction is positive and live load can cause a negative reaction, the load combination would be $0.9DC + 0.65DW + 1.75(LL+IM)$. If both reactions were negative, the load combination would be $1.25DC + 1.50DW + 1.75(LL+IM)$. For each force effect, both extreme combinations may need to be investigated by applying either the high or the low load factor as appropriate. The algebraic sums of these products are the total force effects for which the bridge and its components should be designed.

Water load and friction are included in all strength load combinations at their respective nominal values.

For creep and shrinkage, the specified nominal values should be used. For friction, settlement and water loads, both minimum and maximum values need to be investigated to produce extreme load combinations.

Section 3 -Loads and Load Factors (SI)

Table 3.4.1-1 Load Combinations and Load Factors

Load Combination Limit State	DC DD DW EH EV ES	LL IM CE BR PL LS	WA	WS	WL	FR	TU CR SH	TG	SE	EQ	IC	CT	CV	SC
STRENGTH-I	γ_P	1.75	1.00	-	-	1.00	0.50/1.20	γ_{TG}	γ_{SE}	-	-	-	-	-
STRENGTH-II	γ_P	1.35	1.00	-	-	1.00	0.50/1.20	γ_{TG}	γ_{SE}	-	-	-	-	-
STRENGTH-III	γ_P	-	1.00	1.40	-	1.00	0.50/1.20	γ_{TG}	γ_{SE}	-	-	-	-	-
STRENGTH-IV EH, EV, ES, DW, DC ONLY	γ_P 1.5	-	1.00	-	-	1.00	0.50/1.20	–	–	-	-	-	-	-
STRENGTH-V	γ_P	1.00	1.00	1.20	1.20	1.00	0.50/1.20	γ_{TG}	γ_{SE}	-	-	-	-	-
EXTREME EVENT -I	γ_P	0.25	1.00	-	-	1.00	-	–	–	1.00	-	-	-	-
EXTREME EVENT-II	γ_P	0.25	1.00	0.30	-	1.00	-	–	–	-	1.00	1.00	1.00	-
EXTREME EVENT-III	γ_P	1.75	1.00	-	-	1.00	-	γ_{TG}	γ_{SE}	-	-	-	-	1.80
EXTREME EVENT-IV	γ_P	-	1.00	1.40	-	1.00	-	γ_{TG}	γ_{SE}	-	-	-	-	0.70
EXTREME EVENT-V	γ_P	-	1.00	-	-	1.00	-	-	-	-	1.00	1.00	1.00	0.60
EXTREME EVENT-VI	γ_P	-	1.00	-	-	1.00	-	-	-	1.00	-	-	-	0.25
SERVICE-I	1.00	1.00	1.00	0.30	0.30	1.00	1.00/1.20	γ_{TG}	γ_{SE}	-	-	-	-	-
SERVICE-II	1.00	1.30	1.00	-	-	1.00	1.00/1.20	-	-	-	-	-	-	-
SERVICE-III	1.00	0.80	1.00	-	-	1.00	1.00/1.20	γ_{TG}	γ_{SE}	-	-	-	-	-
FATIGUE – LL, IM & CE ONLY	-	0.75	-	-	-	-	-	-	-	-	-	-	-	-

Table 3.4.1-2 -Load Factors for Permanent Loads, γ_P

Type of Load	Load Factor	
	Maximum	Minimum
DC: Component and Attachments	1.25	0.90
DD: Downdrag	1.80	0.45
DW: Wearing Surface and Utilities	1.50	0.65
EH: Horizontal Earth Pressure		
• Active	1.50	0.90
• At-rest	1.35	0.90
EV: Vertical Earth Pressure		
• Overall Stability	1.35	N/A
• Retaining Structure	1.35	1.00
• Rigid Buried Structure	1.30	0.90
• Rigid Frames	1.35	0.90
• Flexible Buried Structures other than Metal Box Culverts	1.95	0.90
• Flexible Metal Box Culverts	1.50	0.90
ES: Earth Surcharge	1.50	0.75

Section 3 -Loads and Load Factors (SI)

SPECIFICATIONS

The load factor for temperature gradient, γ_{TG} and settlement, γ_{SE} shall be determined on a project-specific basis.

COMMENTARY

The load factor for temperature gradient should be determined based on:

• the type of structure, and

• the limit state being investigated

At this writing (1994), there is general agreement that in situ measurements of temperature gradients have yielded a realistic distribution of temperatures through the depth of some types of bridges, most notably concrete box girders. There is very little agreement on the significance of the effect of that distribution. It is generally acknowledged that cracking, yielding, creep and other non-linear responses diminish the effects. Therefore, load factors of less than 1.0 should be considered, and there is some basis for lower load factors at the strength and extreme event limit states than at the service limit state.

Similarly, open girder construction and multiple steel box girders have traditionally, but perhaps not necessarily correctly, been designed without consideration of temperature gradient, i.e., γ_{TG} = 0.0.

Past editions of the Standard Specifications used γ_{LL} = 0.0 in combination with earthquake effects. The possibility of partial live load, i.e., $\gamma_{LL} < 1.0$, with earthquake should be considered. Application of the Ferry-Borges Model for combining load events indicates that γ_{LL} = 0.25 is reasonable for a wide range of values of ADTT and a representative set of earthquake intensity and occurrence rate data.

3.4.2 Load Factors for Construction Loads

Load factors for the weight of the structure and appurtenances shall not be taken less than 1.25.

Unless otherwise specified by the Owner, the load factor for construction loads, for equipment and for dynamic effects shall not be less than 1.5. The load factor for wind shall not be less than 1.25. All other load factors shall be taken as 1.0.

3.4.3 Load Factor for Jacking Forces

Unless otherwise specified by the Owner, the design forces for jacking in service shall not be less than 1.3 times the permanent load reaction at the bearing, adjacent to the point of jacking. Where the bridge will not be closed to traffic during the jacking operation, the jacking load shall also contain a live load reaction consistent with the maintenance of traffic plans, multiplied by the load factor for live load.

The design force for post-tensioning anchorage zones shall be taken as 1.2 times the maximum jacking force.

C3.4.2

The load factors presented here should not relieve the contractor from the responsibility for safety and damage control during construction.

C3.4.3

The load factor of 1.2 applied to the maximum tendon jacking force results in a design load of about 96% of the nominal ultimate strength of the tendon. This number compares well with the maximum attainable jacking force, which is limited by anchor efficiency.

Section 3 -Loads and Load Factors (SI)

SPECIFICATIONS

COMMENTARY

C3.8.1.3

Based on practical experience, maximum live loads are not expected to be present on the bridge when the wind velocity exceeds 90 km/hr. Hence, the load factor corresponding to the treatment of wind on structures for the strength limit state is reduced to 1.20 in combination with a live load factor of 1.00. A 0.30 wind load factor is used for Service I limit state. Similarly, a 0.30 wind load factor is used for combinations involving vessel collisions

3.15 SCOUR: SC

The evaluation of scour depths at bridge piers and abutments shall follow the procedure outlined in Hydraulic Engineering Circular No. 18.

C.3.15

The HEC-18 manual (Richardson and Davis, 1995) presents the state of practice for the design, evaluation, and inspection of bridges for scour.

A scour safety factor equal to 2.00 should be used when checking foundation depths with the scour calculated from the HEC-18 model. The 2.00 factor ensures that bridges susceptible to scour have reliability levels similar to those of other bridges.

The scour safety factor is reduced to 1.80 for the combination of scour and live loads.

A scour safety factor equal to 0.60 is used for combinations involving vessel collisions.

A scour safety factor equal to 0.70 is used for combinations involving scour and wind loads.

A scour safety factor equal to 0.25 is used for combinations involving scour and earthquakes. Because the presence of scour may result in the reduction of bridge stiffness and thus a reduction in the inertial forces, it is critical that bridge safety be checked for earthquake effects with and without the presence of scour.

Section 3 - Loads and Load Factors (SI)

REFERENCES

Nowak A.S. (1999), NCHRP Report 368: Calibration of LRFD Bridge Design Code, NCHRP 12-33, Transportation Research Board, Washington, DC.

Ghosn, M., Moses, F. & Wang, J. (2003), NCHRP Report 489: Design of Highway Bridges for Extreme Events, NCHRP Project 12-48, Transportation Research Board of the National Academies, Washington, DC.

Ghosn, M. and Moses, F. (1998), NCHRP Report 406: Redundancy in Highway Bridge Superstructures, NCHRP Project 12-36(2), Transportation Research Board of the National Academies, Washington DC.

Liu, D., Ghosn, M., Moses, F., and Neuenhoffer, A. (2001), NCHRP 458: Redundancy in Highway Bridge Substructures, NCHRP Project 12-47, Transportation Research Board of the National Academies, Washington DC.

Richardson, E.V., and Davis, E.R. (1995), Evaluating Scour at Bridges, HEC-18, Federal Highway Administration, FHWA-IP-90-017, Washington, DC.

Section 2 -General Design and Location Features (SI)

SPECIFICATIONS

COMMENTARY

2.6.4.4.2 Bridge Scour

As required by Article 3.7.5, scour at bridge foundations is investigated for two conditions:

- The design flood for scour: The streambed material in the scour prism above the <u>factored total scour line</u> shall be assumed to have been removed for design conditions. The design flood storm surge tide, or mixed population flood shall be the more severe of the 100-year event or an overtopping flood of lesser recurrence interval. <u>Appropriate load combination and scour factors are provided in Table 3.4.1-1 under Extreme Events III, IV, V and VI.</u>
- The check flood for scour: The stability of bridge foundation shall be investigated for scour conditions resulting from a designated flood storm surge tide, or mixed population flood <u>equal to or exceeding the design flood but</u>, not to exceed the 500-year event or an overtopping flood of lesser recurrence interval. <u>A scour factor of 2.0 is specified for checking foundation stability under this condition.</u>

If the site conditions, due to ice or debris jams, and low tailwater conditions near stream confluences dictate the use of a more severe flood event for either the design or check flood for scour, the Engineer may use such flood event.

Spread footings on soil or erodible rock shall be located so that the bottom of footing is below the <u>factored scour depths</u> determined for the check flood for scour. Spread footings on scour-resistant rock shall be designed and constructed to maintain the integrity of the supporting rock.

C2.6.4.4.2

A majority of bridges that have failed in the Unite States and elsewhere have failed due to scour.

The added cost of making a bridge less vulnerable to damage from scour is small in comparison to the to the cost of a bridge failure.

The design flood for scour shall be determined on the basis of the Engineer's judgment of the hydrologic and hydraulic flow conditions at the site. The recommended procedure is to evaluate scour due to the specified flood flows and to design the foundation for the event expected to cause the deepest <u>factored total scour</u>.

The recommended procedure for determining the <u>factored total scour depth</u> at bridge foundations is as follows:

- estimate the long-term channel profile aggradation or degradation over the service life of the bridge,
- estimate the long-term channel plan form change over the service life of the bridge,
- as a design check, adjust the existing channel and flood plain cross-sections upstream and downstream of bridge as necessary to reflect anticipated changes in the channel profile and plan form,
- determine the combination of existing or likely future conditions and flood events that might be expected to result in the deepest <u>factored scour</u> for design conditions,
- determine water surface profiles for a stream reach that extends both upstream and downstream of the

Section 2 -General Design and Location Features (SI)

SPECIFICATIONS

Deep foundations with footings shall be designed to place the top of the footing below the estimated <u>factored contraction scour depth</u> where practical to minimize obstruction to flood flows and resulting local scour. Even lower elevations should be considered for pile supported footings where the piles could be damaged by erosion and corrosion from exposure to stream currents. Where conditions dictate a need to construct the top of a footing to an elevation above the streambed, attention shall be given to the scour potential of the design.

When tendering or other pier protection systems are used, their effect on pier scour and collection of debris shall be taken into consideration in the design.

COMMENTARY

bridge site for the various combinations of conditions and events under consideration,

- determine the magnitude of the <u>factored contraction scour and local scour</u> at piers and abutments,
- evaluate the results of the scour analysis, taking into account the variables in the methods used, the available information on the behavior of the watercourse, and the performance of existing structures during past floods. Also consider present and anticipated future flow patterns in the channel and its flood plain. Visualize the effect of the bridge on these flow patterns and the effect of the flow on the bridge. Modify the bridge design where necessary to satisfy concerns raised by the scour analysis and the evaluation of the channel plan form.

Foundation designs should be based on the <u>factored total scour depths</u> estimated by the above procedure, taking into account appropriate geotechnical safety factors <u>and load combination factors</u>. Where necessary, bridge modifications may include:

- relocation or redesign of piers or abutments to avoid areas of deep scour or overlapping scour holes from adjacent foundation elements,
- addition of guide banks, dikes or other river training works to provide for smoother flow transitions or to control lateral movement of the channel,
- enlargement of the waterway area, or
- relocation of the crossing to avoid an undesirable location.

Foundations should be designed to withstand the conditions of scour for the design flood and the check flood. In general, this will result in deep foundations. The design of the foundations of existing bridges that are being rehabilitated should consider underpinning if scour indicates the need. Riprap and other scour countermeasures may be appropriate if underpinning is not cost effective.

Available technology has not developed sufficiently to provide reliable scour estimates for some conditions, such as bridge abutments, located in areas of turbulence due to converging or diverging flows.

The stability of abutments in areas of turbulent flow shall be thoroughly investigated and exposed embankment slopes should be protected with appropriate scour countermeasures.

124

Section 3 –Loads and Load Factors (SI)

SPECIFICATIONS

3.7.5 Change in Foundations Due to Limit State for Scour

The provisions of Article 2.6.4.4 shall apply.

The consequences of changes in foundation conditions of loaded structures resulting from the design flood for scour shall be considered using the factored scour depth and load combination factors provided in Table 3.4.1-1 for service limit states and extreme event limit states III, IV, V and VI.

Foundation instability due to factored scour depths resulting from the check flood for bridge scour and from hurricanes shall be considered using a factored scour depth. For the check of foundation stability a factor of 2.0 is specified.

COMMENTARY

C3.7.5

Statistically speaking, scour is the most common reason for the failure of highway bridges in the United States.

Provisions concerning the effects of scour are given in Section 2. Scour per se is not a force effect, but by changing the conditions of the substructure it may significantly alter the consequences of force effects acting on structures.

APPENDIX B
RELIABILITY MODEL FOR SCOUR ANALYSIS

Scour is typically assessed using the methodologies presented in FHWA's Hydraulic Engineering Circular No. 18 (HEC-18) (Richardson and Davis, 1995), although many other methods exist. The equations in HEC-18 are deterministic; they do not account for uncertainties in the models, the model parameters, or the hydraulic and hydrologic variables. Pier scour is given in HEC-18 by

$$\frac{y_s}{y_1} = 2.0 K_1 K_2 K_3 K_4 \left(\frac{b^-}{y_1}\right)^{0.65} Fr_1^{0.43} \tag{1}$$

where

$$
\begin{aligned}
y_s &= \text{scour depth,} \\
y_1 &= \text{the upstream flow depth,} \\
b\text{N} &= \text{the effective pier width,} \\
Fr &= \text{the Froude number, and} \\
K_1, K_2, K_3, \text{and } K_4 &= \text{correction factors for the pier shape,} \\
&\quad \text{angle of attack, bed forms, and sediment gradation, respectively.}
\end{aligned}
$$

Adjustments to the pier width are given for the case of piles and pile caps or footings. Each of the four correction factors is given in tables and equations provided in HEC-18.

Monte Carlo simulation can be used to generate random samples of the parameters in Equation 1 based on the specified coefficients of variation and distributions. Equation 1 can be modified by a model correction factor λ to account for uncertainty in the model form and coefficients (Ang and Tang, 1984) as follows:

$$\frac{y_s}{y_1} = 2.0 \lambda K_1 K_2 K_3 K_4 \left(\frac{b^-}{y_1}\right)^{0.65} Fr_1^{0.43} \tag{2}$$

The selection of coefficients of variation and distributions are dependent on each bridge and on difficulties in assessing parameters at that specific bridge. To obtain a probabilistic scour depth, the following steps are followed:

1. Values for each of the random variables in Equation 2, including the model correction factor, are generated from their respective distributions for each of N simulation cycles.
2. Pier scour is calculated from Equation 2 based on the generated random variables.
3. This process is repeated for N simulation cycles.
4. The mean and standard deviation are calculated for the N values of scour depth.

5. A distribution is determined based on the N values of scour depth.

The following example, taken from HEC-18, is used to generate probabilistic scour depths for 5-, 10-, 25-, 50-, and 100-year hydrologic events. The means and coefficients of variations for the variables in Equation 2 are given in Table 1 for the range of hydrologic events (in terms of T-year return periods). The coefficient of variation for flow velocity was calculated based on Manning's resistance equation and assuming that n and S were the only significant sources of uncertainty. Manning's equation is given by

$$V = \frac{\phi}{n} R^{2/3} S^{1/2} \tag{3}$$

where

$$
\begin{aligned}
n &= \text{Manning's roughness coefficient,} \\
R &= \text{hydraulic radius, and} \\
S &= \text{slope.}
\end{aligned}
$$

The uncertainty in V based on the uncertainty in n and S is given as follows (Mays and Tung, 1992):

$$\Omega_V^2 = \Omega_n^2 + 0.25 \Omega_S^2 \tag{4}$$

where Ω is the coefficient of variation. Assuming that flow depth is determined from the standard step method, the uncertainty in flow depth can be calculated based on the results of uncertainty analyses conducted by the U.S. Army Corps of Engineers (Hydraulic Engineering Center, 1986):

$$\Omega_y = 0.76 y^{0.6} S^{0.11} (5 N_r)^{0.65} \tag{5}$$

where N_r is the reliability estimate for n, $0 \leq N_r \leq 1$. In this example, it is assumed that $N_r = 0.5$ (moderate reliability). Uncertainty in K_3 is assumed. Uncertainty in λ is based on comparisons of observed scour depths for 515 sites around the world with calculated values from Equation 1 (Johnson, 1995).

Using the procedure outlined above, the data in Table 1, and 1,000 simulation cycles, probabilistic scour depths were calculated. Table 2 provides the deterministic scour depths computed for each return period based on the HEC-18 equation (Equation 1). The 1,000 scour depths for each return period yield a normal distribution (Johnson and Dock, 1998) with mean $\bar{y}_s$ and standard deviation S_{ys}, given in Table 2.

126

TABLE 1 Parameter estimates, coefficients of variation, and distributions for a hypothetical bridge and a range of return periods

Variable	Probability Distribution	5-year		10-year		25-year		50-year		100-year	
		Mean	Ω	Mean	Ω	Mean	Ω	Mean	Ω	Mean	Ω
b (m)	N/A	1.52	0.000	1.52	0.000	1.52	0.000	1.52	0.000	1.52	0.000
V (m/s)	symmetrical triangular	2.04	0.280	2.29	0.28	2.68	0.28	2.91	0.28	3.73	0.28
y (m)	symmetrical triangular	1.14	0.075	1.37	0.084	1.73	0.097	1.95	0.10	2.84	0.13
λ	asymmetrical triangular	0.55	0.520	0.55	0.520	0.55	0.520	0.55	0.520	0.55	0.520
K_1	N/A	1.0	0.000	1.0	0.000	1.0	0.000	1.0	0.000	1.0	0.000
K_2	N/A	1.0	0.000	1.0	0.000	1.0	0.000	1.0	0.000	1.0	0.000
K_3	uniform	1.1	0.050	1.1	0.050	1.1	0.050	1.1	0.050	1.1	1.000
K_4	N/A	1.0	0.000	1.0	0.000	1.0	0.000	1.0	0.000	1.0	0.000

TABLE 2 Computed scour depths using HEC-18 equation (Equation 1), simulated scour depths (including correction factor λ), and standard deviations

Return Period (years)	Computed Scour Depth (m) from Equation 1	Simulated Mean Scour Depth (m) from Equation 2	Standard Deviation (m)
5	2.53	1.39	0.46
10	2.71	1.49	0.48
25	3.00	1.65	0.49
50	3.11	1.71	0.47
100	3.58	2.07	0.48

Thus, for each return period the probability of exceeding scour depths ranging from 0.5 m to 5.5 m can be calculated as follows:

$$p(y_s \geq k) = p\left(z \geq \frac{k - \bar{y}_s}{S_{y_s}}\right) \qquad (6)$$

where

p = probability,
y_s = scour depth,
k = selected scour depth,

z = standard normal variate,
$\bar{y}_s$ = mean scour depth, and
S_{ys} = standard deviation of scour depth.

Table 3 shows these results. From Tables 2 and 3, the probability of exceeding the scour depths given by HEC-18 can be estimated. For example, for the 100-year return period, the HEC-18 equation (Equation 1) yields a value of 3.6 m. The probability of exceeding this scour depth is approximately 0.0013. Such a low exceedance probability is to be expected given that the HEC-18 equation is intended to predict a conservative, maximum scour depth.

TABLE 3   Exceedance probabilities for selected scour depths

Selected Scour Depth, k (m)	5-year	10-year	25-year	50-year	100-year
0.5	0.9735	0.9804	0.9905	0.9950	0.9993
1	0.8017	0.8463	0.9077	0.9346	0.9848
1.5	0.4055	0.4917	0.6202	0.6725	0.8705
2	0.0924	0.1440	0.2375	0.2686	0.5364
2.5	0.0079	0.0177	0.0414	0.0464	0.1721
3	2.327E-04	0.0008	0.0029	0.0030	0.0237
3.5	2.251E-06	1.411E-05	7.986E-05	6.992E-05	0.0013
4	6.998E-09	8.530E-08	8.106E-07	5.521E-07	2.475E-5
4.5	6.897E-12	1.803E-10	3.017E-09	1.463E-09	1.743E-7
5	2.109E-15	1.321E-13	4.075E-12	1.288E-12	4.335E-10
5.5	0	0	0	0	3.773E-13

REFERENCES FOR APPENDIX B

Ang, A.H-S., and Tang, W.H. (1984). *Probability Concepts in Engineering Planning and Design, Volume II*. Wiley and Sons, NY.

Hydraulic Engineering Center (1986). *Accuracy of Computed Water Surface Profiles*. U.S. Army Corps of Engineers, Davis, CA.

Johnson, P.A. (1995). Comparison of Pier Scour Equations Using Field Data. *Journal of Hydraulic Engineering,* ASCE, 121(8); pp. 626–629.

Johnson, P.A., and Dock, D.A. (1998). Probabilistic Bridge Scour Estimates. *Journal of Hydraulic Engineering,* ASCE, 124(7); pp. 750–755.

Mays, L.W., and Tung, Y.K. (1992). *Hydrosystems Engineering and Management*. McGraw-Hill, Inc.

Richardson, E.V., and Davis, S.R. (1995). *Evaluating Scour at Bridges*, 3rd edition. Report No. FHWA-IP-90-017, Hydraulic Engineering Circular No. 18, Federal Highway Administration, Washington, DC.

APPENDIX C

RELIABILITY ANALYSIS OF THREE-SPAN BRIDGE MODEL

1. TRADITIONAL SCOUR EVALUATION APPROACH

The AASHTO LRFD specifications (1994) state that "a majority of bridges that have failed in the United States and elsewhere have failed due to scour." This is confirmed by Shirole and Holt (1991), who observed that over the last 30 years, more than 1,000 of the 600,000 U.S. bridges have failed and that 60% of these failures are due to scour while earthquakes accounted for only 2%. Of course, there are many more bridges that are posted or otherwise taken out of service due to their inadequate strengths (e.g., due to deterioration, low rating, fatigue damage, etc.); nevertheless, scour is considered a critical cause of failure because its occurrence often leads to total collapse. For these reasons, developing methods for the design and maintenance of bridge foundations for scour is currently considered a top priority for agencies concerned with the safety of bridges.

The AASHTO LRFD specifications require that scour at bridge foundations be designed for the 100-year flood storm surge tide or for the overtopping flood of lesser recurrence interval. The corresponding 100-year design scour depth at bridge foundations is determined following the procedure recommended by FHWA, using what is known as "HEC-18" (Hydraulic Engineering Circular No. 18 [Richardson and Davis, 1995]). The foundation should then be designed taken into consideration the design scour depth. This is achieved by, for example, placing the footings below the scour depth, ensuring that the lengths of piles and pile shafts extend beyond the scour depth, and verifying that the remaining soil depth after scour provide sufficient resistances against shear failures and overturning.

HEC-18 recognizes that the total scour at a highway crossing is comprised of three components:

1. Long-term aggradation and degradation,
2. Contraction scour, and
3. Local scour.

Aggradation and degradation are long-term elevation changes in the streambed of the river or waterway caused by erosion and deposition of material. Contraction scour is due to the removal of material from the bed and the banks of a channel, often caused by the bridge embankments encroaching onto the main channel.

Local scour involves the removal of material from around bridge piers and abutments. It is caused by an acceleration of flow around the bridge foundation that accompanies a rise in water levels that may be due to floods and other events. Local scour and contraction scour can be either clear-water or live-bed. Live-bed conditions occur when there is a transport of bed material in the approach reach. Clear-water conditions occur when there is no bed material transport. Live-bed local scour is cyclic in nature as it allows the scour hole that develops during the rising stage of the water flow to refill during the falling stage. Clear-water scour is permanent because it does not allow for a refill of the hole.

In this research effort, attention was focused on local live-bed scour around bridge piers that, because of its cyclical nature, is the most unpredictable type of scour. For local scour around bridge piers, HEC-18 recommends the use of the following design equation to predict the 100-year design scour depth:

$$y_{max} = 2y_0 K_1 K_2 K_3 K_4 \left(\frac{D}{y_0}\right)^{0.65} F_0^{0.43} \qquad (1)$$

where

$$\begin{aligned} y_{max} &= \text{the maximum depth of scour;} \\ Y_0 &= \text{the depth of flow just upstream of the} \\ &\quad \text{bridge pier excluding local scour;} \\ K_1, K_2, K_3 \text{ and } K_4 &= \text{coefficients that account for the nose} \\ &\quad \text{shape of the pier, the angle between} \\ &\quad \text{the direction of the flow and the direc-} \\ &\quad \text{tion of the pier, the streambed condi-} \\ &\quad \text{tions, and the bed material size;} \\ D &= \text{the pier diameter, and} \\ F_0 &= \text{the Froude number defined as follows:} \end{aligned}$$

$$F_0 = \frac{V}{(gy_0)^{0.5}} \qquad (2)$$

where V is the mean flow velocity at the pier, and g is the acceleration due to gravity.

The process that is followed to calculate the 100-year design (or nominal) scour depth for a bridge pier is as follows:

1. Use statistical data on flood events for the bridge site to obtain the expected maximum 100-year flood discharge rate.
2. Perform a hydraulic analysis to obtain the corresponding expected maximum 100-year flow velocity, V, and the 100-year stream flow depth, y_0.
3. Use the information from Step 2 to calculate the Froude number.

4. Use information on pier geometry, streambed conditions and angle of attack, and streambed material to calculate D, K_1, K_2, K_3 and K_4.
5. Substitute the values obtained into the HEC-18 Equation 1 to calculate the 100-year design scour depth for the bridge under consideration.

The objective of this project, NCHRP Project 12-48, is to calibrate load factors for the scour extreme event and the combination of scour with other extreme events such as earthquakes, winds, and ship collisions. The calibration should be performed using reliability methods to be compatible with the AASHTO LRFD specifications. The calibration process requires the evaluation of the reliability inherent in current design practice. The steps of the calibration process involve

1. "Designing" a number of typical bridges to satisfy current specifications.
2. Finding the inherent reliability in these designs.
3. Finally, calibrating a set of load factors to ensure that bridges designed with these factors will satisfy a target reliability that will be determined based on the experience gained from previous designs.

This appendix focuses on developing a model for calibrating load factors for bridges subjected to the combination of scour and earthquake forces. Current design practice for scour uses the HEC-18 equations published by FHWA. This section illustrates the use of HEC-18 for finding the scour depth at a bridge pier. The methods will subsequently be used to develop a model to study the reliability of bridges subjected to scour and a model to study the reliability of bridges subjected to the combined effects of scour and earthquake forces.

1.1 Illustrative Example

To illustrate the use of the HEC-18 equation to find the scour depth at a bridge pier, we assume that we are to design a bridge for the 100-year scour. The bridge is to be located over a 220-ft wide segment of a river, and the bridge will have the configuration shown in Figure 1.

The bridge is constructed over a 200-ft wide river. To obtain realistic results for different possible discharge rates, data from different small size rivers are used and design scour depths are calculated for each of these river discharge rates. The rivers chosen for this analysis consist of the following: (1) Schohaire Creek in upstate New York, (2) Mohawk River in upstate New York, (3) Cuyahoga River in northern Ohio, (4) Rocky River in Ohio, and (5) Sandusky River in Ohio. Data on the peak annual discharge rates for each of the five rivers was obtained from the U.S. Geological Survey (USGS) website. Lognormal probability plots and Kolmogorov–Smirnov (K-S) goodness-of-fit tests showed that the peak annual discharge rate, Q, for all five rivers can be well modeled by lognormal probability distributions. The mean of the log Q and its standard deviation were calculated using a maximum likelihood estimator. This data are provided in Table 1, along with the K-S maximum difference in cumulative distribution, Dn. D_n^* is the K-S maximum difference between the measured cumulative probability and expected probability value. More than 60 data points were available for each of the five rivers. This indicates

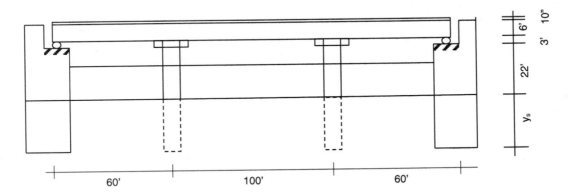

Figure 1. Profile of example bridge.

TABLE 1 Probability models for five rivers

River	Log Q	$S_{log\ Q}$	D_n^*
Schohaire	9.925	0.578	0.067
Mohawk	9.832	0.243	0.068
Sandusky	9.631	0.372	0.086
Cuyahoga	9.108	0.328	0.065
Rocky	9.012	0.378	0.049

that the lognormal distribution is acceptable for a significant levels $\alpha = 20$ percent.

In this example, we will assume that the width of the channel at the bridge site is permanently set at 220 ft, as shown in Figure 1. The rate of flow, Q, at a given point in time is related to the cross sectional area of the stream, A_0, and the stream flow velocity, V, by

$$Q = A_0 V. \tag{3}$$

For a rectangular cross section with a constant width, b, and flow depth, Y_0, the relationship becomes

$$Q = b y_0 V. \tag{4}$$

The flow velocity, V, depends on the flow depth, y_0. The relationship between the flow velocity, V, and flow depth, y_0, can be expressed using Manning's equation:

$$V = \frac{\Phi}{n} R^{2/3} S^{1/2} \tag{5}$$

where

n = the Manning roughness coefficient, which would vary between 0.025 to 0.035 for earth (respectively for good condition and for weeds and stones);
R = the hydraulic radius;
S = the slope of the bed stream; and
Φ = an empirical factor equal to 1.486 when using U.S. units (ft and sec).

For SI units, $F = 1.0$. For the problem described in Figure 1, the hydraulic radius, R, can be calculated by

$$R = \frac{b y_0}{2 + y_0}. \tag{6}$$

In this illustration, we will assume a slope of $S = 0.2\%$. Using the equations given above, the relationship between the flow rate Q and the flow depth y is given from the equation

$$Q = b y_0 \frac{\Phi}{n} \left(\frac{b y_0}{b + 2 y_0} \right)^{2/3} S^{1/2} \tag{7}$$

where the following is input data:

$b = 220$ ft,
$\Phi = 1.486$,
$n = 0.025$, and
$S = 0.2\%$.

For the discharge rate data for each river shown in Table 1, the flow depth for the 100-year flood can be calculated as y_0 along with the corresponding velocity v. Given this information and the geometric data of the river and the bridge pier, the maximum design scour depth can be calculated from Equation 1. In this example, it is assumed that the round nose pier is aligned with the flow and that the bed material is sand. HEC-18 then recommends that the factors K_1, K_2, and K_4 all be set equal to 1.0. For plane bed condition, K_3 is equal to 1.1. Using Equation 1 with the column diameter of $D = 6$ ft, the design scour depth is obtained for each river as shown in Table 2.

1.2 Discussion

To study the safety of bridges accounting for the combined effect of scour and other extreme events, we need to know the extent of scour for different flood intensities, and we also need to know how the scour depth varies with time. This includes the time it takes for the flood to produce the maximum scour and the duration of the foundation exposure after the occurrence of scour before refill. Time becomes an important parameter because it controls the probability of having a simultaneous occurrence of scour and other events.

The HEC-18 approach has been used extensively for practical design considerations although the HEC-18 empirical model provides conservative estimates of scour depths and is known to have the following limitations:

1. The HEC-18 equation is based on model scale experiments in sand. In a recent evaluation against full-scale observations from 56 bridge sites, it was found to vastly over-predict the scour depth (Landers and Mueller, 1996). A comparison between the HEC-18 equation and the measured depths are illustrated in Figure 2, which is adapted from Landers and Mueller (1996).

TABLE 2   Design scour depth for each river

River	Q 100-year (ft³/sec)	V (ft/sec)	Y_0-flood depth (ft)	Y_{max}-scour depth (ft)
Schohaire	78,146	17.81	20.56	17.34
Mohawk	32,747	12.87	11.78	13.99
Sandusky	36,103	13.35	12.52	14.33
Cuyahoga	19,299	10.5	8.45	12.26
Rocky	19,693	10.58	8.56	12.32

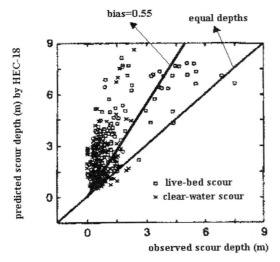

Figure 2. Comparison of HEC-18 predictions with observed scour depths.

2. Once a flood begins, it takes a certain amount of time for the full extent of erosion to take effect. Thus, if the flood is of a short duration, the maximum scour depth may not be reached before the flood recedes. On the other hand, prior floods may have caused partial erosions accelerating the attainment of the maximum scour depth. HEC-18 does not predict the length of time required for the maximum scour depth to be reached and assumes that the maximum depth is always reached independent of the flood duration and the level of scour incurred by prior floods.

3. The HEC-18 model does not distinguish between live-bed scour and clear-water scour in terms of the time required to reach equilibrium scour depth and the differences in the expected magnitudes of scour depths for these different phenomena (Richardson and Davis, 1995).

4. HEC-18 was developed based on experimental data obtained for sand materials. Some work is in progress to extend the use of HEC-18 for both sand and clay streambed materials although it is known that these materials behave differently (Briaud et al., 1999).

5. The usual assumption is that scour is deepest near the peak of a flood, but may be hardly visible after floodwaters recede and scour holes refill with sediment. However, there are no known methods to model how long it takes a river to back-fill the scour hole. Refill can occur only under live-bed conditions and depends on the type and size of the transported bed material (i.e., sand or clay). In any case, even if refill occurs, it will take a considerable time for the refill material to consolidate enough to restore the pier foundation to its initial strength capacity. Although such information is not precisely available, a number of bridge engineers have suggested that periods of 2 to 3 months are reasonable for clay materials with longer periods required for sands.

6. Based on the observations made above, it is evident that using the 100-year flow velocity and flow depth in the HEC-18 equations does not imply that the annual probability of failure will be 1/100. This is primarily due to the inherent conservative bias associated with the HEC-18 equation. In fact, based on observed scour depths at 515 sites, Johnson (1995) has shown that the HEC-18 equation produces scour depth estimates that are on the average 1.8 times higher than the measured values. This shows that the HEC-18 equation has a bias equal to 0.55 (1/1.8). The observed differences between predictions and measurements produced a coefficient of variation (COV; standard deviation/mean) of 52%. Figure 2 shows that these estimates are reasonable when compared with the data collected by Landers and Muller (1996).

Based on the observations made herein, an appropriate reliability model for the safety analysis of a bridge pier under the effect of scour is proposed, as illustrated in Section 2 of this report. This model is based on the work presented in Appendix B of the interim report for NCHRP Project 12-48.

2. RELIABILITY ANALYSIS FOR SCOUR

The HEC-18 model stipulates that previous levels of scour at the site do not affect the scour depth produced by a given flood. Hence, the maximum scour in a given return period is a function of the maximum flood observed in that period and is not affected by previous smaller floods that may have occurred within that same period. In addition, the HEC-18 model assumes that the flood duration is always long enough for the full scour depth corresponding to the flood velocity to be reached.

Although the scour hole is normally assumed to refill as the scour-causing flood recedes, the available literature does not provide precise information on how long it normally takes for the foundation to regain its original strength. This is believed to depend on the type of material being deposited by live-bed streams. For example, cohesive materials such as clays may tend to regain their strength within a short period of time (perhaps 2 to 3 months). On the other hand, fine sands may take much longer to consolidate and regain their original strengths. As a compromise, for this illustrative example, we will assume that it will take about 6 months (½ year) for a foundation to regain its original strength. We will assume that the scour depth produced by the maximum yearly flood will remain at its maximum value for this half-year period. This assumption will also indirectly account for the effects of smaller floods within that period of time. In the proposed model, we will also assume that the scour depth will be reached instantaneously as the flood occurs and that the flooding period is always long enough for the maximum scour depth to be reached.

Using these assumptions, the proposed model for occurrence of scour can be represented as shown in Figure 3. T is

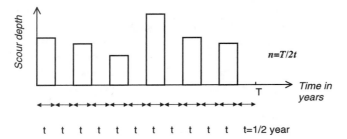

Figure 3. Representation of proposed scour model.

the lifetime of the bridge (e.g., 75 years), t is equal to the duration of the maximum scour ($1/2$ year) for each time unit (1 year). n is the number of scour occurrences within the time period T (or number of scour repetitions). For the 75-year design life of a bridge, $n = 75$, assuming one flooding season per year.

The model represented in Figure 3 can be used to calculate the reliability of a foundation subjected to scour alone or to the combination of scour and other extreme events. The reliability calculations for scour alone are illustrated in Section 2.1 in this appendix. The reliability calculations for the combination of scour and earthquake loads are illustrated in Section 5. In order to illustrate the load combination problem, a reliability model for earthquake alone is needed. Such a model is presented in Section 4.

2.1 Reliability Analysis for Scour Alone

The model used herein for the reliability analysis of a bridge foundation subjected to scour alone is based on the example provided in Appendix B of the interim report for NCHRP Project 12-48. The calculations use the following assumptions:

1. The probability distribution for the maximum 1-year flood is expressed as $F_Q(x)$, which gives the cumulative probability that the 1-year discharge volume, Q, is less than x. The cumulative probability distribution of the maximum 1-year discharge volume Q follows a lognormal distribution. This is based on probability plots and K-S goodness-of-fit tests.

2. A 75-year design life (return period) is assumed in order to be consistent with the AASHTO LRFD specifications.

3. The maximum 75-year flood discharge has a cumulative probability distribution $F_{Q75}(x)$ related to the probability distribution of the 1-year maximum discharge by

$$F_{Q75}(x) = F_Q(x)^{75}. \qquad (8)$$

Equation 8 assumes independence between the floods observed in different years. This assumption is consistent with current methods to predict maximum floods.

4. For a specific discharge rate Q, the corresponding values for the flow velocity V and flow depth y_0 can be calculated by solving Equation 7 for y_0 and then using Equation 5 to find V.

5. It is noted that determining the appropriate Manning roughness coefficient (n in Equations 5 and 7) is associated with a high level of uncertainty. Therefore, in this example, we will assume that n is a random variable with a mean value equal to 0.025 (the recommended value for smooth earth surfaces) and a coefficient of variation equal to 28%. The 28% COV was adapted from the Hydraulic Engineering Center (1986), which recommends that n follow a lognormal distribution with a COV ranging from 20% to 35% (with an average value of 28%). It is herein assumed that the slope S is known and, thus, the uncertainties in V are primarily due to the uncertainties in determining n.

6. For any set of values for V and y_0, one can calculate the Froude number from Equation 2 and substitute these values into Equation 1 to find the 100-year design scour depth. Equation 1, however, as explained earlier, gives a safe value for the depth of scour. The average value was found by Johnson (1995) to be about 0.55 times the nominal value (bias value is 0.55). Also, the ratio of the true scour value over the predicted value has a COV of 52%. This bias is represented by a modeling parameter λ_{sc}. Thus, the true scour value is given by an equation of the form

$$y_{\max} = 2\lambda_{sc} y_0 K_1 K_1 K_1 K_1 \left(\frac{D}{y_0} \right)^{0.65} F_0^{0.43} \qquad (9)$$

7. It should be noted that Johnson (1995) also recommends that the factor K_3, representing the effect of streambed condition, be treated as random variable with a bias equal to 1.0 and a COV equal to 5%.

8. Data on the type of the probability distributions for λ_{sc} and K_3 are not available. In the interim report for this project, Johnson proposed to use triangular distributions. It is herein assumed that n will follow a lognormal distribution as recommended by the Hydraulic Engineering Center (1986). This will ensure that n does not take values less than 0. On the other hand, because it is unlikely that λ_{sc} and K_3 will have upper or lower bounds, we shall herein assume that these two random variables follow normal (Gaussian) distributions. Note that triangular distributions where assumed in the interim report for both these variables. A sensitivity analysis will be performed at a later stage of this project to study how these assumptions will affect the final results.

The information provided above can be used in a simulation program to find the probability that the scour depth in a 75-year period will exceed a given value y_{cr}. A summary of the input data used in a Monte Carlo simulation program that calculates the probability that the actual scour depth will exceed a critical value is shown in Table 3.

The probability that y_{max} will exceed a critical scour depth, y_{cr}, is calculated for different values of y_{cr} as shown in Figure 4 for all five rivers. This data are also summarized in Table 4. The results of the simulation can be summarized as shown in Table 4 for the five rivers. It is observed that the safety index implied in current scour design procedures is on the order of 1.40 to 1.50, which is much lower than the 3.5 safety index used as the basis for the calibration of the load factors for the combination of live and dead loads.

Table 5 gives the load factor that should be applied on the scour design equation in order to produce different values of reliability indexes. For example, if a target reliability index of $\beta_{target} = 3.50$ is desired, then the average load factor that should be used in designing bridge foundations for scour should be equal to 1.69. Similarly, if a reliability index $\beta_{target} = 2.50$ is to be used as the target index for the design of bridge

foundations for scour, then the load factor should be equal to 1.32. This means that a "load" (safety factor) equal to 1.69 to satisfy a target index of 3.5 or equal to 1.32 to satisfy a target index of 2.50 should further multiply the scour depth obtained from Equation 1. The load factors for other target safety index values ranging from 4.0 to 2.0 are provided in Table 5.

2.2 Final Remarks

The reliability calculations shown above assumed a deterministic value of y_{cr}. In general the critical depth, y_{cr}, that will produce the failure of the foundation is a random variable that depends on the foundation type, soil properties, and the type of loading (lateral or vertical) applied on the structure. For example, under the effect of lateral loads, pile foundations may fail in shear or in bending. Probability of failure would increase in the presence of scour that extends the moment arm of the lateral load. In addition, scour would decrease the depth of the remaining soil available to resist the applied moment and shear forces. On the other hand, when a dynamic analysis is performed, it is observed that the presence of scour reduces the stiffness of the bridge foundation

TABLE 3 Input data for reliability analysis for scour alone

Variable	Mean value	COV	Distribution Type
Q—maximum 75-yr discharge rate	As provided in Table 1	As provided in Table 1	Lognormal
λ_Q modeling variable for Q	1	5%	Normal
n Manning roughness	0.025	28%	Lognormal
λ_{sc} modeling variable	0.55	52%	Normal
K_3 Bed condition factor	1.1	5%	Normal

Probability of scour depth exceedance

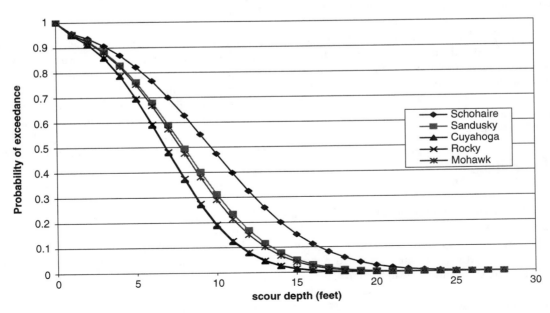

Figure 4. Probability that actual scour will exceed critical depths.

134

TABLE 4   Summary of simulation results

River	Average Q_{75yr} (Q for 75 years) (ft3/sec)	COV of Q_{75yr}	Average y_{s75} (max.scour depth in 75 yrs)	COV of Y_{s75}	Design depth (ft)	Reliability index β
Schohaire	85,000	29%	9.8	52%	17.3	1.40
Mohawk	34,000	12%	7.8	51%	14.0	1.51
Sandusky	38,000	18%	8.0	51%	14.3	1.41
Cuyahoga	20,000	16%	6.9	51%	12.3	1.42
Rocky	21,000	19%	6.9	51%	12.3	1.40

TABLE 5   Design scour depths required to satisfy target reliability levels

Target index Required depth (ft) River	β = 4.0 Depth	Load factor	β = 3.5 Depth	Load factor	β = 3.0 Depth	Load factor	β = 2.5 Depth	Load factor	β = 2.0 Depth	Load factor
Schohaire	33.5	1.90	30.0	1.73	26.0	1.50	23.0	1.33	20.0	1.15
Mohawk	26.0	1.86	23.0	1.64	20.5	1.46	18.5	1.32	16.0	1.14
Sandusky	26.5	1.85	24.0	1.68	21.0	1.47	19.0	1.33	17.0	1.19
Cuyahoga	22.0	1.79	20.5	1.67	18.0	1.46	16.0	1.30	14.0	1.14
Rocky	23.0	1.87	21.0	1.71	18.0	1.46	16.0	1.30	14.0	1.14
Average		*1.85*		*1.69*		*1.46*		*1.32*		*1.15*

thus creating a more flexible system that often leads to a reduction in dynamic loads. The example solved in the next section of this report will illustrate how the lateral load will affect the safety of bridges for different failure modes.

The controlling random variable for the reliability analysis is the modeling factor λ_{sc}, which has a high COV of 52%. This COV is higher than that of the maximum 75-year flood, which is about 15%. It should be noted that the analysis performed for scour alone produces a reliability index and probability of failure values that are totally independent of the time period during which the scour hole is at its maximum depth. The effect of the scour duration will be important as we study the reliability of bridge piers due to the combined effects of scour and other loads. Section 3 will analyze the bridge described in Figure 1 for earthquake loads using current analysis methods. Section 4 will develop a model for the reliability analysis of bridge piers subjected to earthquake loads alone. Section 5 will demonstrate the reliability analysis for the combination of earthquake loads and scour.

3. ANALYSIS OF BRIDGE FOR EARTHQUAKE LOADS

In this appendix, we are illustrating the case in which scour is to be combined with seismic forces. In order to study the combination problem, we first perform a deterministic analysis of the bridge under the effect of seismic forces to determine the nominal design forces and moments. In Section 4, the reliability of the column subjected to seismic forces alone is executed. In Section 5, the reliability analysis is expanded to demonstrate how the combination of scour and earthquakes will affect the reliability of the column and the foundation. The reliability analysis will be used to develop load factors that will be calibrated to achieve consistent target reliability levels. A first attempt on the calibration is presented in Section 6.

3.1 Bridge Geometric Properties

The bridge example studied in this section consists of three spans (60 ft, 100 ft, 60 ft) as illustrated in Figure 1 of Section 1.1. The bridge has two bents, each of which is formed by a 6-ft-diameter column. The weight applied on each bent is calculated to be 1,527 kips divided as follows: superstructure weight = 979 kips, weight of substructure = 322 kips, and weight of wearing surface = 2.46 kip/ft. Following current practice, the tributary length for each column is 91.9 ft where 50 ft is 50% of the distance between the columns and 41.9 ft is 70% of the distance between the column and the external support. This assumes that the lateral connection of the superstructure to the abutments will not break because of the earthquake lateral motions. The clear distance between the base of the column and the center of the superstructure is 25 ft. The foundation consists of a pile shaft (pile extension) that extends 50 ft into the soil. The soil is assumed to have an elastic modulus of Es = 10,000 psi corresponding to moderately stiff sand. The point of fixity of the floating foundation can be calculated using the relationship provided by Poulos and Davis (1980), given as follows:

$$\left(\frac{L_e}{L}\right)^3 + 1.5\frac{e}{L}\left(\frac{L_e}{L}\right)^2 = 3K_R\left(I_{\rho H} + \frac{e}{L}I_{\rho M}\right) \quad (10)$$

where

L_e = the effective depth of the foundation (distance from ground level to point of fixity);

L = the actual depth;

e = the clear distance of the column above ground level;

K_R = the pile flexibility factor which gives the relative stiffness of the pile and soil;

$I_{\rho H}$ = the influence coefficient for lateral force; and

$I_{\rho M}$ = the influence coefficient for moment.

The pile flexibility factor is given as follows:

$$K_R = \frac{E_p I_p}{E_s L^4}. \tag{11}$$

If the pile is made of 4,000 psi concrete, then E_p equaling 3600 ksi (= 57[4000]$^{1/2}$) and the diameter of the column being $D = 6$ ft results in a moment of inertia $I_p = 63.62$ ft^4 (= $\pi r^4/4$). Thus, for a pile length of 50 ft, the pile flexibility becomes

$$K_R = \frac{E_p I_p}{E_s L^4} = \frac{3,600,000 \times 63.64}{10,000 \times 50^4} = 0.037. \tag{11'}$$

The charts provided by Poulos and Davis (1980) show that for $K_R = 0.0037$, the influence coefficients $I_{\rho H}$ and $I_{\rho M}$ are respectively on the order of 5 and 15. This will produce an equation of the following form:

$$\left(\frac{L_e}{50}\right)^3 + 1.5\frac{25}{50}\left(\frac{L_e}{50}\right)^2 = 3 \times 0.0037 \times \left(5 + \frac{25}{50}50\right) \tag{10'}$$

where the value of $e = 25$ ft, which is the clear height of the column.

The root of this equation produces a ratio of $L_e/L = 0.35$, resulting in an effective depth of 18 ft below ground surface. Thus the effective total column height until the point of fixity becomes 25 ft + 18 ft = 43 ft.

For transverse seismic motion, the bent is assumed to be fixed at the base of the effective pile depth and free on the top. Thus, the bent stiffness is

$$K_{bent} = \frac{3EI}{H^3} \tag{12}$$

where

H = the effective column height ($H = e + L_e$),

E = the column's modulus of elasticity, and

I = the moment of inertia.

For typical concrete columns, E equals 3,600 ksi. For a circular column with a diameter of $D = 6$ ft, the moment of inertia is

$$I = \frac{\pi r^4}{4} = \frac{\pi 3^4}{4} = 63.64 \, ft^4. \tag{13}$$

For a height equal to 45 ft, Equation 13 gives the bent stiffness as follows:

$$K_{bent} = \frac{3EI}{H^3} = \frac{3 \times 3600 \text{k/in}^2 \times 144 \text{ft}^2/\text{in}^2 \times 63.64 \text{ft}^4}{(43 \text{ft})^3} \tag{12'}$$

$$= 1244 \text{ kip/ft}.$$

The natural period of the bent system, T is

$$T = \frac{2\pi}{\omega} = 2\pi\sqrt{\frac{M}{K}} = 2\pi\sqrt{\frac{1527 \text{ kip}/32.2 \text{ ft/sec}^2}{1244 \text{ kip/ft}}} \tag{14}$$

$$= 1.23 \text{ sec}$$

where

ω = the circular frequency of the system;

M = the mass ($M = W/g$ where W is the weight and g is the acceleration of gravity); and

K = the stiffness.

According the AASHTO LRFD specifications, the elastic seismic response coefficient C is given by

$$C = \frac{1.2AS}{T^{2/3}} \leq 2.5A \tag{15}$$

where

A = the acceleration coefficient (the seismic acceleration as a multiple of the acceleration of gravity),

S = the soil type parameter, and

T = the natural period of the system.

The AASHTO LRFD specifications (Figure 3.10.2-2 of the manual) provide maps that give seismic acceleration for different regions of the United States. For example, for bridges to be located in the New York City area, the map of the acceleration coefficients shows a value of A on the order of 15% to 18% of g. Current trends in earthquake design of highway bridges are favoring the use of the National Earthquake Hazards Reduction Program (NEHRP) spectra, and a new earthquake design specification for bridges is proposing to use the same NEHRP spectra developed for buildings. For this reason, this appendix is based on using the NEHRP spectra rather than the current AASHTO specifications. The sections below illustrate the application of the NEHRP approach for the design of the bridge illustrated in Figure 1 of Section 1.1.

3.2 Earthquake Intensity

The USGS mapping project gives the peak ground accelerations (PGAs) at bedrock level and the corresponding spectral values for 10%, 5%, and 2% probability of exceedance in 50 years for points throughout the United States. The spectral accelerations are given for periods of 0.2, 0.3, and 1.0 sec corresponding to the three different PGAs. Table 6 lists these values for the areas with the following zip codes: 10031 in New York City, 38101 in Memphis, 55418 in St. Paul, 98195 in Seattle, and 94117 in San Francisco. The spectral accelerations give the accelerations of the mass of a single-degree-of-freedom system that is supported by a system that has a natural period of 0.2, 0.3, or 1.0 sec when the acceleration at the base of the system is equal to the corresponding PGA. The values shown in Table 6 are in %g where g is the acceleration due to gravity. Table 6 shows the accelerations for three different probabilities of exceedance, namely 10% in 50 years, 5% in 50 years, and 2% in 50 years. For example, the 2% in 50 years corresponds to an earthquake return period of 2,500 years. The 10% in 50 years corresponds to a return period of 500 years. These return periods are normally used as the bases of current bridge design practice.

3.3 NEHRP Earthquake Response Spectrum

The spectral accelerations provided in Table 6 are for single-degree-of-freedom (SDOF) systems founded on bedrock. The values are given for only three natural periods—0.2, 0.3,

TABLE 6   Probabilistic ground motion values, in %g, for five sites (based on USGS website)

Site	10% PE in 50 yr	5% PE in 50 yr	2% PE in 50 yr
S.Francisco			
PGA	52.65	65	76.52
0.2 sec SA	121.61	140.14	181
0.3 sec SA	120.94	140.44	181.97
1.0 sec SA	57.7	71.83	100.14
Seattle			
PGA	33.77	48.61	76.49
0.2 sec SA	75.2	113.63	161.34
0.3 sec SA	62.25	103.36	145.47
1.0 sec SA	22.06	32.23	55.97
St. Paul			
PGA	0.76	1.31	2.5
0.2 sec SA	1.82	3.17	5.63
0.3 sec SA	1.61	2.72	4.98
1.0 sec SA	0.73	1.38	2.66
New York			
PGA	6.32	11.92	24.45
0.2 sec SA	12.59	22.98	42.55
0.3 sec SA	9.42	16.64	31.17
1.0 sec SA	2.85	5.11	9.4
Memphis			
PGA	13.92	30.17	69.03
0.2 sec SA	27.46	58.71	130.03
0.3 sec SA	20.38	43.36	110.62
1.0 sec SA	6.46	15.47	40.74

and 1.0 sec—in addition to the PGA. NEHRP has proposed a method to use the information provided in Table 6 to develop acceleration response spectra that are valid for various soil conditions and for systems with different natural periods. The NEHRP response spectra can be described by a curve with the shape shown in Figure 5. In Figure 5, S_a is the spectral acceleration, T is the natural period of the system, S_{Ds} is the maximum spectral acceleration, and S_{D1} is the spectral acceleration for a period of $T = 1$ sec. All spectral accelerations are given as function of g, the acceleration due to gravity. T_0 gives the period at which the maximum spectral acceleration is reached. T_s gives the period at which the spectral acceleration decreases below the maximum value. When the period T is less than T_0, the spectral acceleration increases linearly. When the period T is greater than T_s, the spectral acceleration is inversely proportional to T. The values of S_{Ds} and S_{D1}, as well as T_0 and T_s, are calculated from the spectral accelerations given in Table 6 and the soil properties as described further below.

The first step in the earthquake analysis process is to develop the spectral response curve of Figure 5. This requires the identification of the foundation soil type. Having a modulus of elasticity on the order of 10,000 psi would classify the site condition as stiff soil (NEHRP Soil Category D). This information will be used to obtain the site coefficients (or site amplification factors).

For a 500-year return period (10%PE in 50 years), the mapped spectral acceleration, S_s, for the short period of $T = 0.2$ sec is taken from Table 6 for each of the five sites. Similarly, the spectral acceleration S_1 for a period of $T = 1$ sec is obtained from Table 6 for each site. The site coefficients are obtained from the NEHRP provisions as F_a for short periods and F_v for 1 sec from Table 4.1.2.4.a and 4.1.2.4.b of NEHRP (1997). Thus, the maximum earthquake spectral response accelerations for short period (0.2 sec) S_{MS} and for the 1-sec period S_{M1} adjusted for the proper soil profile are obtained from

$$S_{MS} = F_a S_S \quad \text{and} \quad S_{M1} = F_v S_1. \quad (16)$$

NEHRP allows for a $^2/_3$ ($= 0.667$) correction on the maximum earthquake accelerations. Thus, the critical amplitudes on the response spectrum are

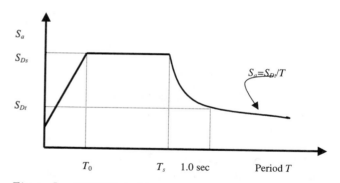

Figure 5.   NEHRP design response spectrum.

$$S_{MS} = 0.667 S_{MS} = 0.667 FaS_S \qquad \text{and}$$

$$S_{M1} = 0.667 S_{M1} = 0.667 F_V S_1. \tag{17}$$

All spectral values are given in function of the acceleration due to gravity, g. Critical periods on the response spectrum (see Figure 5) are

$$T_0 = 0.20 S_{D1}/S_{DS} \qquad \text{and} \qquad T_s = S_{D1}/S_{DS} \tag{18}$$

The equation describing the acceleration response spectrum, S_a, shown in Figure 5 will be

$$S_a = 0.6 \frac{S_{DS}}{T_0} T + 0.40 S_{DS} \qquad \text{for } T < T_0$$

$$S_a = S_{DS} \qquad \text{for } T_0 < T < T_s \tag{19}$$

$$S_a = \frac{S_{D1}}{T} \qquad \text{for } T > T_s$$

For the 2,500-year return period (2%PE in 50 years), the mapped maximum earthquake spectral response for short periods, S_s, are also taken from Table 6, and the determination of the spectral curve is executed following Equations 16 through 19 for each of the five sites. The site coefficients, F_a and F_v (site amplification factors), are also obtained from the NEHRP provisions. The maximum earthquake spectral response accelerations adjusted for the proper soil profile are given in Table 7 for the five sites studied for a natural period of $T = 1.23$ sec, which is the period of the system studied in this example.

3.4 Design Moment Capacity of Column

Using a response modification factor of $R = 2.0$ for single columns of essential structures, the applied design forces and moments are reduced by $R = 2.0$ to account for column ductility. The equivalent force on the top of the column is obtained by multiplying the weight by the spectral acceleration (S_a) corresponding to the natural period of $T = 1.23$ sec. The moment at the base of the pile shaft is obtained by multiplying the equivalent force by the effective column height H. Thus, the final design moment in the column is obtained as follows:

$$M_{design} = \frac{WS_a H}{R}. \tag{20}$$

The weight applied on the bridge column W is 1,527 kips. The spectral acceleration values are taken from Table 7. The effective height of the column, H, includes the clear height plus the depth to the point of fixity ($H = e + L_e = 25$ ft + 18 ft = 43 ft). Using Equation 20, the design moments for the column are calculated as given in Table 8.

3.5 Check of Foundation Safety

The design of the foundation should be such that the soil must be able to resist either the elastic forces applied on the bridge system or the forces produced from the plastic hinging of the column. The smaller of the two resulting forces is used for designing the depth of the bridge foundation, implying that if the earthquake forces induce the formation of the plastic hinge in the column, then only the forces transmitted by the hinge need to be considered for the soil resistance. It is only in the case in which the columns are so overdesigned that no hinges will form in them that the soil needs to carry the full elastic forces applied on the bridge system.

For piles in cohesionless soils, the depth of the pile bent should be such that the soil pressure resists the applied lateral forces. Using the Broms model as described by Poulos and Davis (1980), the lateral load capacity of the soil is given by

$$H_u = \frac{0.5 \gamma D L^3 K_p}{e + L} \tag{21}$$

where

γ = the soil's specific weight;
L = the depth of the foundation;
D = the diameter of the foundation;
e = the column eccentricity above ground level; and
K_p = Rankine's passive pressure coefficient, given by

$$K_p = \frac{1 + \sin\phi}{1 - \sin\phi} \tag{22}$$

where ϕ is the angle of internal friction of the soil.

TABLE 7   Spectral accelerations, S_a, for 10% and 2% PE in 50 years for bridge column (based on NEHRP specifications)

Site	10%PE in 50 yr	2%PE in 50 yr
San Francisco	0.469	0.814
Seattle	0.234	0.455
St. Paul	0.009	0.035
New York	0.037	0.122
Memphis	0.084	0.352

TABLE 8   Design moments for 10% and 2% PE in 50 years (based on NEHRP specifications)

	10% PE in 50 yr (kip-ft)	2% PE in 50 yr (kip-ft)
San Francisco	15397.5	26724.03
Seattle	7682.34	14937.88
St. Paul	295.47	1149.07
New York	1214.73	4005.32
Memphis	2757.76	11556.34

138

For design purposes, the applied lateral force resulting from the formation of the plastic hinge in the bridge column is obtained from the over strength moment capacity of the column, which is 1.3 times the plastic moment capacity (i.e., 1.3 times the values shown in Table 8).

If the clear column height is $e = 25$ ft and the effective depth to the point of fixity is $L_e = 18$ ft with $H = 43$ ft ($H = e + L_e$), then the equivalent shearing force produced in the hinge accounting for the column's over strength will be

$$H_{\text{applied}} = \frac{1.3M_p}{H} \qquad (23)$$

where M_p is taken from Table 8. Using this information, the required foundation depth for each site is given as shown in Table 9.

Notice that the calculations of the effective depth and the applied moment performed above during the determination of the column design capacity assumed a foundation depth of 50 ft. The results of Table 9 show that the 50-ft depth is conservative for most bridge sites. It is also noted that the point of fixity for the bridge/foundation system under lateral loads is on the order of 16 to 18 ft for all foundations exceeding 30 ft in actual depth. An iterative process may be used to find the optimum foundation depth as a function of site information. In addition to resisting the lateral load, the foundation depth should be such that the pile be able to carry the applied vertical loads due to the dead weight of the structure and the live load expected during its design life. The required foundation depth to resist the vertical loads is calculated in Section 3.6. The reliability calculations performed in Section 4 studies the effect of different column depths on the safety of bridge columns against the failure of the column in bending and the bearing capacity of the foundation.

3.6 Safety Check of Foundation for Vertical Loads

In addition to carrying the horizontal load produced by the earthquake tremors, the foundation should be able to carry the vertical loads applied on the structures including the dead weight of the superstructure and substructure as well as the

TABLE 9 Required foundation depth based on column capacity

Site	Required foundation depth (ft)	
	10% PE in 50 yr	2% PE in 50 yr
San Francisco	34.7	43.7
Seattle	26.2	34.3
St. Paul	7.6	12.5
New York	12.8	20.2
Memphis	17.5	30.8

vertical live load. Piles resist the vertical loads caused by the combination of friction resistance and bearing resistance.

The friction and bearing resistances are a function of the pile dimensions and soil properties. In the case studied herein, the angle of friction for sandy soil is taken as $\phi = 35°$, the modulus of elasticity as $E_s = 10,000$ psi, the Poisson ratio as $\nu = 0.3$, the unit weight as $\gamma = 60$ lb/ft^3, and the pile diameter as $D = 6$ ft. Poulos and Davis (1980) show that the vertical stress in the soil will vary linearly up to a level of $z = 7$ times the pile diameter, $z = 7D$. This means that the stress will vary linearly up to 42 ft below ground level, after which point the stress remains constant. For saturated soils, the maximum pressure will be due to the combination of water and soil weights, thus the maximum stress will be $\sigma_v = 42$ ft $\times$ 60 lb/ft^3 = 2,520 lb/ft^2. For bored piles with $\phi = 35°$, the bearing capacity factor N_q is given by Poulos and Davis (1980) as $N_q = 40$. Thus, the bearing capacity of the pile, P_{bu}, at the 42-ft level and below will be as follows:

$$P_{bu} = A_b \sigma_v N_q \qquad (24)$$

where A_b is the pile base diameter. For $A_b = 28.27$ ft^2 ($= \pi D^2/4$), $\sigma_v = 2,520$ lb/ft^2, and $N_q = 40$, the bearing capacity of the pile at the 42-ft depth and below is $P_{bu} = 2.85 \times 10^6$ lb or 2,850 kips. The friction force, P_{su}, for up to z ft in depth is given by

$$P_{su} = \frac{\pi D(\sigma_v K_s \tan\phi)z}{2}. \qquad (25)$$

The value of $K_s \tan\phi = 0.20$ is provided by Poulos and Davis for a friction angle $\phi = 35°$ and for bored piles. Thus, the friction force P_{su} for a depth of $z = 42$ ft will be equal to 200 kips. After the 42-ft depth, the friction force will be a constant function of $L - 42$ ft. This means that the total friction force will be 200 kips + 9.50 ($L - 42$).

The combination of P_{su} and P_{bu} should be able to carry the applied vertical loads from the superstructure as well as the weight of the pile. Thus, the final bearing capacity P_u is given as

$$P_u = 2,850 \text{ kips} + 200 \text{ kips} + 9.50 \text{ kip/ft} (L - 42 \text{ ft}). \qquad (26)$$

Given a pile of length, L, the weight of the pile will be 150 lb/ft$^3 \times L \times \pi D^2/4$. The applied weight of the structure had been earlier given as $W = 1,527$ kips.

According to the AASHTO LRFD specifications (1994), the live load applied on a continuous bridge is due to 90% of the AASHTO lane load plus 90% of the effects of two AASHTO design trucks. The reaction at the interior support due to these loads are on the order of 350 kips. Using the LRFD equation with a dead load factor $\gamma_d = 1.25$ and a live load factor $\gamma_l = 1.75$ results in a required bearing capacity resistance of

$$\phi P_{\text{req.}} = \gamma_d 1,527 + \gamma_l 350 = 2,520 \text{ kips.} \qquad (27)$$

For the bearing capacity of piles in sandy soils, the AASHTO LRFD does not provide a resistance factor ϕ, although values on the order of $\phi = 0.50$ to 0.65 are recommended for clay soils depending on the models used for calculating the bearing capacity. A factor of $\phi = 0.80$ is used when the pile bearing capacity is verified from load tests. By comparing Equations 26 and 27, it is clear that the AASHTO LRFD criteria can be met for a pile length $L = 50$ ft only if pile tests are conducted to verify the bearing capacity of the pile shaft or when a bell at the bottom of the caisson is provided to extend the area. For this example, we are assuming that the 50-ft pile length is acceptable to carry the vertical loads. A sensitivity analysis is performed in the next section to study the effect of different pile lengths on the reliability of the bridge when subjected to earthquake loads.

4. RELIABILITY ANALYSIS OF BRIDGE FOR EARTHQUAKES

The purpose of the analysis performed in this section is to calculate the probability that the bridge designed in Section 3 for the different moment capacities will fail under the effect of earthquake forces within its intended 75-year design life. A 75-year return period is chosen in order to remain compatible with the AASHTO LRFD specifications. The corresponding reliability index, β, will also be calculated. The objective is to study the reliability of bridges designed following current practice and the NEHRP specifications under the effect of earthquake loads. In this section, the analysis is performed assuming no scour. The model is subsequently used in Section 5 to study the reliability of bridges under the combined effects of scour and earthquake loads.

In this section, we will assume that the bridge was designed as described in the previous section to withstand the earthquake loads observed at the five different sites identified above. In order to perform the reliability analysis, we will need to account for the uncertainties associated with each of the random variables that control the safety of the bridge. Assuming no scour, the random variables are identified as

1. Strength capacity of bridge column,
2. Intensity of the earthquake acceleration at the site,
3. Natural period of the column,
4. Mass applied on the column,
5. Seismic response coefficient, and
6. Response modification factor.

In addition, although not treated as a random variable, the frequency of earthquakes at the bridge site plays an important role in determining the reliability of the bridge for earthquake risk. Other factors such as the height and diameter of the column and other geometric and material parameters are associated with very small uncertainties and may be treated as deterministic values.

4.1 Column Strength Capacity

A single-column bridge pier subjected to earthquake loads can fail in a variety of modes. These include (1) failure due to the applied bending moment exceeding the moment capacity of the column (this mode should consider the interaction between bending and axial loads); (2) failure in shear; and (3) failure of the foundations. In this example, the reliability analysis procedure is illustrated for the failure of the column due to the applied bending moment.

The bridge columns under combined axial load and bending moment are normally designed such that the design point at the interaction curve remains below the balance point. In this example, we are ignoring the effects of vertical accelerations as is common in practical cases for short- to medium-length bridges. Hence, the uncertainty in evaluating the vertical load is small compared with that associated with determining the lateral forces (and bending moments). For this reason, we shall assume that the effects of the axial load on the uncertainties in determining the moment capacity are negligible.

According to Nowak (1999), the moment capacity of concrete members in bending is on the average higher than the nominal capacity by a factor of 14% (bias 1.14), and the standard deviation is 13%. The probability distribution of the moment capacity is taken to be lognormal (Nowak, 1999). The final values used in this example are those recommended by Nowak.

4.2 Frequency of Earthquakes

We shall assume that the bridge under consideration may be located in any of the five cities listed in Table 6. The USGS Seismic Hazard Mapping Project stipulates an earthquake occurrence rate for each of the areas. These values are then used by USGS to determine the maximum annual probability curve for earthquake intensities that are discussed in the next section.

4.3 Intensity of Earthquake Accelerations

Frequencies at which earthquakes exceed certain levels of PGAs are provided by the USGS in two different formats: (1) plots showing the probability of exceedance of maximum yearly earthquakes versus PGAs for a number of representative U.S. sites, and (2) tables giving the probability of exceeding given acceleration levels in 50 years. The plots are shown in Figure 6.

The frequency of exceedance in 50 years is related to the maximum yearly earthquake levels, as shown below. For example, if the frequency of exceeding an earthquake level of $0.53g$ is given as 10% in 50 years, this indicates that the probability that the maximum earthquake level in 50 years will be below $0.53g$ is $1.00 - 0.10 = 0.90$. This can be represented as follows:

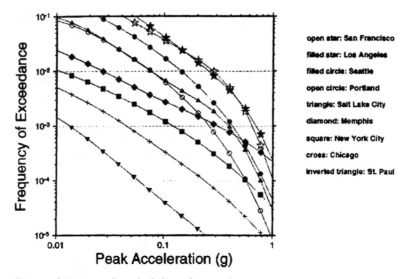

Figure 6. Annual probability of exceedance curves.

$$F_{A50}(0.53g) = 1.00 - 0.10 = 0.90 \qquad (28)$$

where $A50$ is the maximum 50-year acceleration intensity and $F_{A50}(-)$ is the cumulative probability distribution of the maximum 50-year earthquake intensity. But, assuming independence between earthquake intensities (which is a common assumption in earthquake engineering practice), the 50-year probability of exceedance is related to the yearly probability of exceedance, $F_A(x)$, by

$$F_{A50}(x) = F_A(x)^{50}, \qquad (29)$$

hence,

$$F_{A50}(0.53g) = (F_{A50}[0.53g])^{1/50} = 0.997895 \qquad (30)$$

or the yearly probability of exceeding an acceleration of $0.53g$ is 0.2105% ($1 - 0.997895 = 0.002105$). Figure 6 gives the yearly probability of exceedance for a range of accelerations and a number of representative sites. These curves were developed by USGS and available at the USGS website for use in earthquake engineering practice.

In addition to the curves shown in Figure 6, the USGS Earthquake Hazard Mapping Project has provided the project with the data for the New York City, San Francisco, St. Paul, and Seattle sites. These data are entered in the reliability program to represent the earthquake intensity variable. The curve is fitted through a spline function to obtain a cumulative probability distribution that connects the data points. Since bridges are constructed for a 75-year design life, the probability of exceeding different acceleration levels in 75 years will be used in the reliability calculations. These values are obtained using an equation similar to Equation 29 but using an exponent equal to 75 in the right-hand side of the equation. Table 10 gives the values available for the five sites considered.

4.4 Natural Period of the Bridge

The natural period of a bridge depends on many parameters, including the type and the characteristics of the bridge foundation and the stiffness of the soil. Takada et al. (1989) have suggested that the average value of the period is about 1.08 times the value calculated using design methods (bias = 1.08) with a COV on the order of 20%. These values account for the soil structure interaction (SSI) and other analysis effects. The values provided by Takada et al. (1989) are primarily for buildings. The Takada et al. data should be applied on struc-

TABLE 10   Maximum yearly earthquake intensity levels versus probability of exceedance

Ground Motion (g)	Frequency of Exceedance per Year			
	New York	San Francisco	Seattle	St. Paul
0.005	1.74E-02	4.10E-01	2.27E-01	3.27E-03
0.007	1.37E-02	3.67E-01	2.04E-01	2.26E-03
0.0098	1.07E-02	3.17E-01	1.77E-01	1.50E-03
0.0137	8.29E-03	2.63E-01	1.48E-01	9.59E-04
0.0192	6.31E-03	2.06E-01	1.17E-01	5.86E-04
0.0269	4.72E-03	1.53E-01	8.69E-02	3.48E-04
0.0376	3.47E-03	1.09E-01	6.16E-02	2.03E-04
0.0527	2.49E-03	7.55E-02	4.14E-02	1.17E-04
0.0738	1.74E-03	5.11E-02	2.66E-02	6.80E-05
0.103	1.20E-03	3.41E-02	1.63E-02	3.97E-05
0.145	7.88E-04	2.25E-02	9.35E-03	2.29E-05
0.203	5.06E-04	1.50E-02	5.12E-03	1.33E-05
0.284	3.13E-04	9.63E-03	2.64E-03	7.55E-06
0.397	1.86E-04	5.52E-03	1.26E-03	4.17E-06
0.556	1.05E-04	2.44E-03	5.58E-04	2.19E-06
0.778	5.57E-05	7.30E-04	2.39E-04	1.08E-06
1.09	2.78E-05	1.36E-04	9.95E-05	4.95E-07
1.52	1.31E-05	1.67E-05	3.68E-05	2.10E-07
2.13	5.69E-06	1.48E-06	9.85E-06	8.02E-08

tural models that did not include the effects of SSIs as the 1.08 bias accounts for the effects of SSI. Since our analysis model included the effects of SSI, a lower bias should be used.

When SSI models are included in the analysis (as is the case in the models used in this section), the variation between the measured periods and the predicted periods appears to be smaller, and the bias is reduced to a value close to 1.0. This phenomenon is illustrated as shown in Figure 7 adapted from the paper by Stewart et al. (1999). The figure shows that, on the average, the predicted periods accounting for SSI are reasonably similar to the values measured in the field. The bias is found to be 0.99 and the COV on the order of 8.5%. It is noted that the SSI model shown in the figure is different than the model used in this section, and that the comparison is made for buildings rather than bridges. Also, it is noted that the study by Stewart et al. concentrated on the effects of SSI while the natural periods of the structural systems were inferred from field measurements. Thus, the uncertainties in modeling the structures were not included. Furthermore, many of the sites studied by Stewart et al. had relatively stiff soils where the effects of SSI are rather small.

Hwang et al. (1988) report that Haviland (1976) found that the median of natural periods for buildings is equal to 0.91 times the computed values with a COV of 34%. Chopra and Goel (2000) developed formulas for determining the natural periods of buildings based on measured data. The spread in the data shows a COV on the order of 20% for concrete buildings and slightly higher (on the order of 23%) for steel-frame buildings.

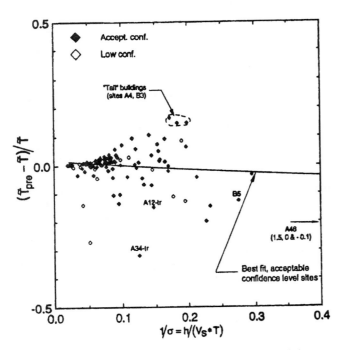

Figure 7. Variation of predicted natural period of the predicted foundation-soil system compared with measured values (from Stewart et al., 1999).

In this appendix, a sensitivity analysis is performed to study the effect of the variability of the natural T on the final results. The analysis shows that the final results are not sensitive to the values used for bias and COV on T. Based on the review of the above references, it is herein decided to use a bias of 0.90 and a COV of 20% for the period of the system. The 0.90 bias similar to that observed by Haviland (1976) is justified based on the fact that the analysis performed to calculate the period of the bridge system used the nominal value for the modulus of elasticity of the concrete, E_c. In reality, the actual modulus of the concrete will be higher than the nominal value and, thus, the predicted stiffness will be larger than the actual stiffness, producing a lower actual period than predicted. A correction factor of 1.20 to 1.30 on the concrete modulus is often used in engineering practice that would justify the 0.90 bias $\left(0.90 \approx 1/\sqrt{1.25}\right)$. The COV of 20% used herein corresponds to the values observed by Chopra and Goel (2000) for buildings. It may be argued that the prediction of the period for bridges may be less uncertain than that for buildings. However, the data collected by Chopra and Goel (2000) did not indicate any appreciable difference in the level of uncertainty due to building heights or sizes. Thus, it may be reasonable to assume that the COV on the period for bridges is also on the order of 20%. The proposed bias and COV are meant to account for both the uncertainties in the structural properties as well as the SSI.

4.5 Mass Applied on the Column

The dead weight effect on bridge members was found by Nowak (1999) to be on the order of 1.05 times the nominal (design) weight with a COV of 10%. This high COV, however, reflects the effects of the structural analysis as well as the uncertainty in estimating the weight. To account for the uncertainty in the weight alone, a COV of 5% is used. The probability distribution is assumed to be normal following the models used by Ellingwood et al. (1980) and Nowak (1999). It is noted that the variability in the mass and the weight considered here are used for calculating the applied forces and not the period. The uncertainty in predicting the mass during the calculation of the period is considered under the biases and COV of T.

4.6 Seismic Response Coefficient

The design spectra proposed for the upcoming AASHTO LRFD for earthquake design of bridge are similar to those adopted by NEHRP. These are based on the average response spectra developed by Frankel et al. (1997) from a large earthquake mapping project. Frankel et al. (1997) found that the level of confidence in NEHRP spectra is related to the number of earthquakes recently observed as well as to knowledge of the type and locations of the faults in a particular region.

142

For sites where a large number of earthquakes was observed, the COV is low; the COV is high for sites with few observed tremors. Frankel et al. provided maps showing uncertainty estimates for selected cities derived from Monte Carlo simulations. The data provided in this map show that the ratio between the 85th fractile and the 15th fractile for New York City is on the order of 5. Assuming a normal distribution, this means that the COV would be on the order of 30%. For the Memphis area, the ratio of the two fractiles is about 3, resulting in a COV of about 25%. For San Francisco, the projected COV is about 15%, and for St. Paul and Seattle the projections are that the COVs would be about 40% and 25%, respectively. Frankel et al. also show that the mean value of the spectral accelerations is very close to the uniform hazard spectra they developed and that resulted in the NEHRP specifications. It is noted that these observations are within the range of the values reported by Seed et al. (1976), who observed that the results of dynamic analyses using a variety of earthquake records resulted in a range of spectral responses with a COV of about 30% from the average spectra.

4.7 Response Modification Factor

The response modification factor is related to the ductility of the system. The purpose of the response modification factor is to allow for a linear elastic analysis of structural systems although the system may exhibit large levels of plastic deformations.

The response modification factor, R, is related to the ductility capacity of the bridge members. Thus, if a member's ductility capacity μ is known, the response modification for that member—assuming an SDOF system—may be evaluated so that the actual plastic response of the structure can be inferred from the linear elastic analysis. Miranda (1997) found that for typical periods of bridge systems (0.5 to 1.5 sec) subjected to a representative sample of earthquake records, the response modification, R, is on the average equal to the bridge column ductility capacity ($R = \mu$) with a COV of about 25%. This observation confirms the model first proposed by Newmark and Hall (1973) that was based on limited data from the El Centro Earthquake. The results of Miranda (1997) were calculated for a variety of sites with a range of soil classifications. Liu et al. (1998), in a report to the National Center for Earthquake Engineering Research (NCEER) and FHWA, found that the COV reduces to about 17% if the earthquake records were chosen to match those that produce the design spectral accelerations.

In addition to the issue of the relation between R and μ, another issue concerns the level of uncertainty associated with estimating the ductility capacity. Results given by Priestly and Park (1987) show that the real ductility of bridge columns are on the average about 1.5 times higher than the ductility estimated from the design formulas with a COV of about 30%. Thus, the actual response modification factor will be on the average 1.5 times the specified ductility capacity, $\mu_{specified}$, (a bias of 1.5), with a COV of 34% $\left(34\% = \sqrt{0.30^2 + 0.17^2}\right)$. The probability distribution for R is assumed to be normal.

The last issue with the response modification factor concerns the range of values specified for use during the design process by AASHTO and other earthquake design codes. For example, it is noted that the response modification factor specified by AASHTO for use during the analysis of single-column bents is set at 2.0, and 3.5 is used for multicolumn bents of essential structures while values of 3.0 and 5.0 are used for "other structures." ATC-6 mentions that an $R = 2.0$ is recommended for a wall-type pier "based on the assumption that a wall pier has low ductility capacity and no redundancy" (Applied Technology Council, 1981). It is clear that the difference among the 2.0, 3.0, 3.5, and 5.0 values of R used for the design of columns is not intended to account for the differences in the ductility capacities of the columns. Rather, the use of different values of R is meant to provide certain types of structures (particularly nonredundant and essential bridges) with higher levels of safety. In fact, since in all cases the design and construction procedures of columns in single-column bents or multicolumn bents are fairly similar, one would expect to find the ductility capacities of all columns to be about the same. It is noted that previous recommendations for the design of bridges under earthquake loads recommended that a response modification factor of $R = 8$ be used. In addition, tests on bridge columns performed at the University of Canterbury (Zahn et al., 1986) have shown that the ductility of properly confined columns can easily exceed 7.5 although some damage would be expected to occur.

On the other hand, the analysis of multicolumn bents produces different moments in each column due to the effect of the dead load and the presence of axial forces. An extensive analysis of different bent configurations founded on different soil types was performed for NCHRP Project 12-47 (Liu et al., 2001). The results showed that, due to the load redistribution and the presence of ductility, multicolumn bents on the average fail at loads up to 30% higher than the loads that make the first column reach its ultimate member capacity.

Based on the information collected from the references mentioned above, we will assume that the nominal ductility level of bridge columns will be equal to $\mu = R = 5.0$. The bias for the ductility level is 1.5 with a COV of 34%. This would results in a mean ductility value of $\mu = R = 7.5$. In addition, multicolumn bents will be associated with a "system overstrength factor" of 1.30 based on the work of Liu et al. (2001). It is noted that the values used herein are similar to those used by Hwang et al. (1988), who have recommended the use of a median value of $R = 7.0$ for shear walls with a COV of 40%.

4.8 Modeling Factor

The structural analysis produces a level of uncertainty in the final estimate of the equivalent applied moment on the base of the bridge column. These factors include the effects of lateral restraints from the slab, the uncertainty in predicting the tributary area for the calculation of mass, the point of application of the equivalent static load, the variability in soil properties and the uncertainty in soil classification, the effect of using a SDOF model, the level of confidence associated with predicting the earthquake intensity, and so forth. Ellingwood et al. (1980) have assumed that the modeling factor has a mean value equal to 1.0 and a COV on the order of 20% for buildings. The same value is used in these calculations.

4.9 Reliability Equation

Using the information provided above, the applied moment on the bridge column is calculated using the following expression:

$$M_{apl} = \lambda_{eq} C' S_a(t'T) * \frac{A * W}{R} * H \tag{30a}$$

where

M_{apl} = the applied moment at the base of the column;
λ_{eq} = the modeling factor for the analysis of earthquake loads on bridges;
C' = the response spectrum modeling parameter;
A = the maximum 75-year PGA at the site (a 75-year design life is used to be consistent with the AASHTO LRFD specifications);
S_a = the calculated spectral acceleration as a function of the actual period;
T = the bridge column period;
t' = the period modeling factor;
W = the weight of the system;
R = the response modification factor; and
H = the column height.

The data used in the reliability analysis for the random variables of Equation 18 are summarized in Table 11. The variables not listed in Table 11 are assumed to be deterministic.

The final safety margin equation for earthquake loads applied on highway bridges can be represented as follows:

$$Z = M_{cap} - M_{apl} \tag{31}$$

where failure occurs if the safety margin Z is less than 0. M_{cap} is the moment strength capacity of the bridge column. Both M_{cap} and M_{apl} are random variables.

4.10 Reliability Results

Using a Monte Carlo simulation along with the safety margin of Equation 31 and the statistical data of Table 11, the reliability index for a 75-year design life of the bridge studied in this report is calculated for different foundation depths. The results illustrated in Figure 8 show that the reliability index is relatively insensitive to the depth of the pile foundation. This is because for the bridge soil type and pile diameter, the point of fixity remains relatively constant at about 16 ft to 18 ft below ground level when the foundation depth exceeds 35 ft. Thus, the stiffness of the system and the moment arm of the column are not affected by the depth of the foundation. Figure 8 also shows that the different sites produced nearly similar values of the reliability index (around 2.50) when the bridge has been designed for the 2% probability of exceedance in 50 years. This confirms that reasonably uniform hazards are achieved for bridge structures designed to satisfy the NEHRP specifications for various sites within the United States.

Figure 9 shows the reliability indexes obtained if the bridge were designed to resist the earthquake with a 10% probability of exceedance in 50 years. By comparing Figures 8 and 9, it is observed that the reliability index decreases to an average of $\beta = 1.90$ from an average of 2.50. Also, it is noted that the range of variation in the reliability index is wider for the lower earthquake level, the range being from about $\beta = 1.6$ to $\beta = 2.1$.

To study the effect of the uncertainties associated with calculating the natural period of the system on the reliability index, the reliability calculations are executed for three different cases with different biases and COVs on T. The

TABLE 11   Summary of input values for seismic reliability analysis

Variable	Mean	Bias	COV	Distribution Type
M_{cap}	From Table 8	1.14	13%	Lognormal
λ_{eq}	1.0	1.0	20%	Normal
C'	1.0	1.0	Varies per site (15% to 40%)	Normal
A	From Table 10	From Table 10	From Table 10	From Table 10
T	1.23 sec	0.9	20%	Normal
W	1527 kips	1.05	5%	Normal
R	5	1.5	34%	Normal

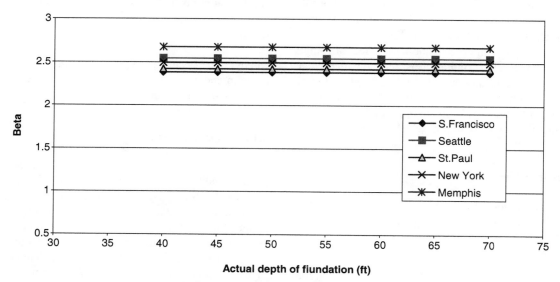

Figure 8. Reliability index versus foundation depth for bridges designed for the earthquakes with 2% probability of exceedance in 50 years.

results executed for the San Francisco site are illustrated in Figure 10. The results show that the final results are relatively insensitive to the variations in the input parameters for *T*. In fact, the calculations show that the uncertainties in determining the maximum earthquake intensity in a 75-year design life are dominant with a COV of 74% for the San Fran-

cisco site and even higher for the other sites (e.g., for New York City, the COV is about 280%). A previously performed first order reliability method (FORM) analysis has shown that the other important factor that affects the reliability index is the response modification factor, *R*, that is associated with a COV of 34%.

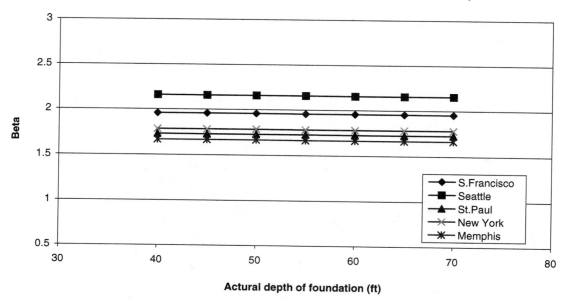

Figure 9. Reliability index versus foundation depth for bridges designed for the earthquakes with 10% probability of exceedance in 50 years.

Reliability index in 10%PE in 50yr in San Francisco (no scour)

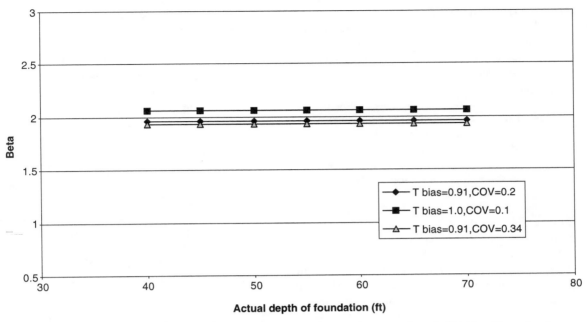

Figure 10.   Sensitivity analysis for different biases and COV on the natural period T *(San Francisco).*

5. COMBINATION OF SCOUR AND EARTHQUAKES

5.1 Monte Carlo Procedure

The analysis of the safety of a bridge subjected to the combination of scour and earthquake loads is a function of the random variables identified in Sections 2 and 4 of this appendix and listed in Tables 3 and 11. Some of these variables vary with time, and others are constant over the design life of the bridge. The reliability analysis involving time-dependent variables in combination with time-independent random variables may be performed based on the Ferry-Borges model for load combination. The logic followed in formulating the combination problem is demonstrated in the following steps.

1. A bridge having the configuration shown in Figure 1 is designed to cross over the 220-ft-wide stream. The stream's characteristics provide a set of variables that are needed for the scour analysis. These characteristics are represented by the following random variables of Equation 9: Manning roughness coefficient, n; the streambed condition factor, K_3; and the modeling factors λ_{sc} and λ_Q. These variables are subject to scatter in their estimation, hence they are considered random. However, these variables remain the same throughout the design life of the bridge. On the other hand, the discharge rate Q is a random variable whose value

changes from one flood to the other. Q is a time-dependent random variable.
2. Similarly, the geometric and other terms affecting the earthquake response of the bridge such as the weight, W, the variation in the estimated period, t', the response modification factor R, the modeling parameters λ_{eq} and the spectral modeling factor C', which are subject to scatter in their estimation and are herein treated as random variables. However, these variables remain the same throughout the design life of the bridge. On the other hand, the earthquake intensity A changes for each earthquake and, thus, like Q, A is also a time-dependent random variable.
3. The Monte Carlo simulation used to calculate the probability of bridge failure in a 75-year design life under the effect of the combination of scour and earthquake loads begins by randomly selecting values for each of the time-independent random variables listed above including the moment capacity (M_{cap}) that is treated as a time-independent random variable with a bias of 1.14 and a COV of 13%.
4. For every flooding season (assumed to be once a year), floods of random discharge intensities will occur in the stream crossed by the bridge described in Figure 1. The maximum annual discharge rate, Q, follows a lognormal distribution with different parameters depending on the river crossed by the bridge as described in Table 1.

5. Given a realization of the random variable Q, the corresponding flow depth, y_0, and velocity, V, are calculated from Equations 7 and 5.

6. For realizations of the modeling factors, λ_{sc} and λ_Q, the roughness factor, n, the bed condition factor, K_3, and the other deterministic parameters, the maximum expected scour depth for the 1 year, y_{max}, is calculated using Equation 9.

7. Since Q is a time-dependent random variable, then y_0 and V are also time-dependent random variables. Consequently, the maximum annual scour depth y_{max} is a time-dependent random variable.

8. The scour depth y_{max} is assumed to remain constant for the duration of the flooding season assumed to be $1/2$ year. During the next flooding season, the value of the scour depth may change depending on the occurrence of the random variable Q.

9. During the $1/2$ year period when y_{max} exists, there is a finite probability that an earthquake will also occur.

10. The probability distribution for the maximum earthquake intensity for this $1/2$-year period can be obtained from the results of Table 10 by using the relationship

$$F_A(x) = F_{Alyr}(x)^{1/2} \qquad (32)$$

where A is the maximum earthquake intensity expected in the $1/2$-year period when scour is present, and A_{lyr} is the maximum earthquake intensity in 1 year as given in Table 10 for each site considered.

11. Given a realization of the scour depth, y_{max} obtained from the scour calculations based on the maximum discharge rate corresponding to 1 year, the effective height of the bridge column becomes

$$H = 25 \text{ ft} + y_{max} + L_e \qquad (33)$$

where 25 ft is the original column height, y_{max} is the scour depth, and L_e is the distance between the ground level and the point of column fixity. The point of fixity is calculated from Equation 10 when the actual depth of the foundation, L, is adjusted by reducing the original depth by y_{max}.

12. The change in column height will produce a change in the stiffness K of Equation 12 and, consequently, will produce a change in the period of the system. (This change would reduce the natural period of the system and thus reduce the earthquake force, but a higher H will also increase the moment arm).

13. On the other hand, the presence of scour will reduce the depth of the foundation by y_{max}. This in turn will change the point of fixity of the column L_e as shown in Equation 10.

14. For the realizations of the time-dependent variables A and y_{max} and all the other time-independent variables, the moment applied on the bridge column is obtained from Equation 30.

15. The moment obtained from Step 14 gives the maximum moment expected to be applied during one flooding season due to the possible combination of earthquake forces and scour.

16. When the flooding season is over, and the scour hole is refilled, an earthquake might still occur and the applied moment for the earthquake that might hit during the dry season is calculated in the same manner described above, but without scour (i.e., with $y_{max} = 0.0$).

17. The higher of the two moments obtained from the dry season and the flooding season becomes the maximum yearly moment for the bridge.

18. Since there are 75 years within the design life of the bridge, the process starting with Step 4 is repeated 75 times. The maximum moment from all the 75 iterations will provide one realization of the maximum 75-year applied moment.

19. Steps 4 through 18 assumed that all the random variables, except for y_{max} and A, are set at values that were kept fixed over the design life of the bridge. These variables are Manning's number, n; the scour modeling factors λ_Q and λ_{sc}; the stream bed condition factor, K_3, as well as the earthquake modeling factor λ_{eq}; the spectral factor, C'; the period, T; the weight, W; and the response modification factor, R. The means, COVs and probability distributions of these variables that are sampled only once for the bridge design life are given in Tables 3 and 11. Failure of the bridge is verified by comparing the maximum 75-year applied moment to the moment capacity established in Step 3.

20. The Monte Carlo simulation is continued by repeating Steps 3 through 19 by first creating another set of realizations for the time-independent random variables, then calculating the maximum 75-year applied moment, and checking whether failure will occur. In the calculations presented in this appendix, the calculations are repeated a total of 100,000 times. The probability of failure is estimated as the ratio of the number of failures that are counted in Step 19 divided by the total number of iterations (100,000). The reliability index, β, is calculated from inverting the following equation:

$$P_f = \Phi(-\beta) \qquad (34)$$

where P_f is the probability of failure and Φ is the cumulative normal distribution function.

21. The reliability calculations use the same safety margin given in Equation 31. This safety margin equation considers only the failure of the bridge column due to bending moment. Other failure modes such as failure in the soil due to the reduction of the soil depth have also been considered as will be discussed further below.

5.2 Results for Bending of Bridge Column

The reliability calculations are first executed for the five earthquake sites with the discharge data of the Cuyahoga River. The results illustrated in Figure 11 show that scour does not significantly affect the reliability index of the bridge for the column failure mode because as explained earlier, the presence of scour reduced the stiffness of the system, but at the same time increased the moment arm resulting in little change in the probability of failure of the column under bending loads. The presence of scour, however, would affect other failure modes, particularly the failure of the soil. This phenomenon is explained in Section 5.3.

5.3 Failure of Soil Due to the Combination of Scour and Earthquake Loads

The bridge could fail in different possible modes. These include moment failure at the base of the columns, shearing failures along the length of the columns, failure of the soil due to lateral loads, and so forth. The analysis performed above has shown that the presence of scour did not significantly affect the safety of the column for failure in bending. In this section, the failure of the foundation due to lateral forces is analyzed.

The model used assumes an equivalent static, linear-elastic behavior of the soil following Rankine's method as described by Poulos and Davis (1980). For the soil conditions of the site, the internal friction angle of the sand, ϕ, is equal to 35°. The buoyant unit weight of sand is 60 lb/ft^3. The free body diagram for the column-foundation-soil system is shown in Figure 12.

In Figure 12, F is the inertial force (equivalent applied force), H gives the effective height of the column to the point of fixity, and P_p is the passive resultant resisting force of the soil (produced by the triangular soil pressure resisting the motion). According to Poulos and Davis (1980), the active force P_p is given as follows:

$$P_p = \frac{3\gamma D K_p L^2}{2} \qquad (34a)$$

where

γ = the specific weight of sand,
L = the depth of the pile;
D = the diameter of the pile, and
K_p = the Rankine coefficient, which is given by

$$K_p = \frac{1 + \sin(\phi)}{1 - \sin(\phi)} \qquad (35)$$

where $\phi = 35°$ is the angle of friction for sand.

5.3.1 Definition of Failure

Failure of the soil due to lateral load may occur under two different cases: (1) no hinges form in the column and the soil is unable to resist the applied lateral force on top of the column and (2) a hinge forms in the column at the point of fixity and the soil is unable to resist the shearing forces transferred from the hinge. For Case 1, the applied lateral force is obtained from Equation 30a as

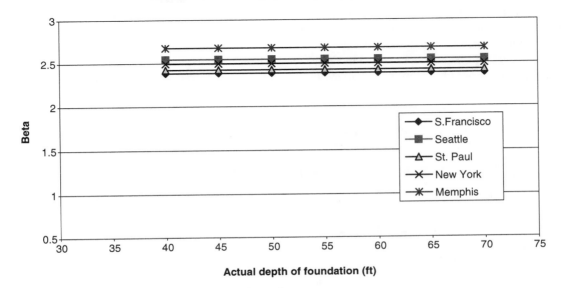

Figure 11. Effect of scour on the reliability index.

148

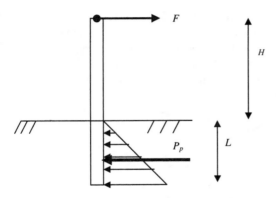

Figure 12.   Free body diagram of bridge system.

$$P_{appl} = \lambda_{eq} C' S_a(t' T) * A * W. \qquad (36)$$

Notice that the response modification factor is not used to calculate the applied lateral force on the system when no hinges are formed (i.e., when the column remains in the linear elastic range).

For Case 2, the applied force is

$$F_{appl} = \frac{M_u}{H} \qquad (37)$$

where M_u is the ultimate moment capacity of the column. The actual ultimate moment capacity is a random variable with statistics related to the nominal capacity through a bias of 1.14 and a COV of 13%, as explained earlier. The safety margin for the failure of the soil is given as

$$Z = P_p - F_{appl} \qquad (38)$$

where failure occurs if the safety margin Z is less than 0. P_p is the lateral force capacity of the soil. The presence of scour will lower the applied force of Equation 36 because it will reduce the stiffness of the system. Scour would increase the moment arm H, which would also lower the applied force

calculated from Equation 37. On the other hand, scour will decrease the length of the foundation L, thus reducing the capacity of the foundation to resist the applied lateral loads. The fact that L is raised to the second power in Equation 34a indicates that the effect of the reduction of L on the foundation's capacity is significant.

Both P_p and F_{appl} of Equation 38 are random variables. Poulos and Davis (1980) mention that the bias observed in test results compared with Equation 34a is on the order of 1.50. In addition, the foundation's capacity to resist lateral load as calculated in Equation 34a is a function of the Rankine pressure coefficient, K_p, the unit weight of sand, γ, and the angle of friction, ϕ. These are random variables with COVs that may exceed 20%. In particular, Becker (1996) gives the following values for the COV for the soil parameters: for the unit weight of soil, $V_\gamma = 7\%$; $V_\phi = 13\%$ for the angle of friction of sand; and for the Rankine pressure coefficients, $V_K = 20\%$. In addition to the uncertainties associated with calculating the Rankine coefficients, the difference between the soil resistance under static loads and dynamic loads must be considered. To account for the dynamic effects on soils, Bea (1983) proposed to use a cyclic factor with a bias of 1.0 and a COV of 15%. In addition to these parameters, which are time-independent random variables, the foundation strength is a function of the dimension of the foundation. The dimensions are all assumed to be deterministic. The summary of the input data used to calculate the reliability of the column foundation to resist lateral load is provided in Table 12.

In addition to the random variables identified in Table 12, the random variables listed in Tables 3 and 11 are used in the Monte Carlo simulation that calculates the reliability of the bridge system subjected to the combination of scour and earthquake loads. The results for the failure of the soil for the San Francisco site are provided in Figure 13.

The results shown in Figure 13 illustrate the importance of the foundation depth on the reliability of the bridge system against soil failure due to lateral loads when the bridge is subjected to the combination of the scour and earthquake loads. The increase in the reliability index is dramatic because a change in the foundation depth from 50 ft to 60 ft produced a change in the reliability index from 2.14 to 3.05 for the case

TABLE 12   Input data for soil-related random variables

Variable	Mean Value	COV	Distribution Type
Unit weight of soil	60 lb/ft^3	7%	Normal
Angle of friction	35°	13%	Normal
Rankine earth pressure	1.5 nominal value	20%	Normal
Cyclic effects	1	15%	Normal

Reliability Index for Soil Failure—Schohaire River Data and San Francisco *EQ* Data

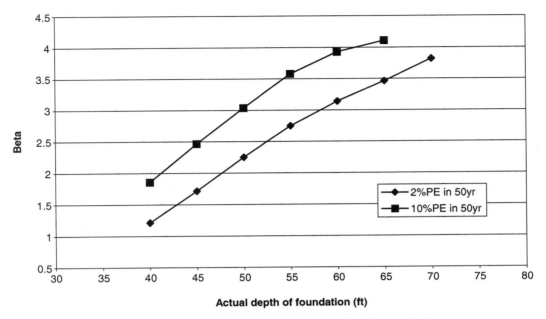

Figure 13. Reliability index for soil failure as a function of foundation depth.

when the bridge is designed such that the column capacity satisfies the NEHRP criteria for 2% probability of exceedance in 50 years. It is also noted that when the bridge columns are designed with weaker moment capacity, as is the case when the design is set to satisfy the earthquake level corresponding to 10% probability of exceedance in 50 years, the reliability index is higher than for 2% case because the columns will transmit a smaller force to the foundation as described by Equation 37. Of course, the lower capacity will mean a higher probability of column failure in bending.

To study the effect of scour on the reliability index for soil failure, the same case analyzed in Figure 13 for the San Francisco earthquake is repeated assuming that no scour could occur. The results are illustrated in Figure 14, which shows that the presence of scour would decrease the reliability index for soil failure by about 1.50 for the case when the foundation is 40-ft deep. The difference decreases to about 0.0 for very deep foundations of 70 ft. This is because the effect of scour on the capacity of the soil is minimal when the foundation is very deep. The results from Figure 14 can be used to calibrate the load factors for the combination of scour and earthquake loads as will be illustrated in Section 6.

6. CALIBRATION OF LOAD FACTORS FOR COMBINATION OF SCOUR AND EARTHQUAKE LOADS

The results obtained in the previous section can be used to calibrate the appropriate load factor for the combination of scour and earthquake loads. For example, as seen in Table 9,

current design specifications would require that the foundation depth be equal to 43.7 ft (rounded up to 44 ft) if the bridge shown in Figure 1 of Section 1.1 were to be located in the San Francisco site. The 44-ft foundation depth would have produced a reliability index of 2.1 if scour were not considered. If the same target reliability index should be attained when scour is included, then the depth of the foundation should be 49 ft. This would require extending the foundation by another 5 ft. Notice that for the Schohaire River (used to model the occurrence of scour in Figure 14), the expected 100-year scour depth is 17.34 ft, as shown in Table 2. Hence, for the bridge to achieve the same target reliability index, it will not be necessary to extend the foundation by the 17.34 ft calculated from the scour alone; an extension of only 5 ft is sufficient. This smaller extension reflects the lower probability of having a high level of scour in combination with a high earthquake load. Also, one should note that the effects of earthquakes and scour are not linear. Hence, the interaction and the correlation between the two effects may require a reduction in the amount of scour considered when the analysis of the bridge for earthquake loads is performed.

It is also noted that if a target reliability index of 3.5 is to be achieved, the foundation depth should be extended to about 63 ft if no scour is to be considered, while the foundation should be about 65-ft deep for the combination of scour and earthquake loads. To satisfy a target reliability index of 2.5, the depth of the foundation for no scour should be 48 ft while 53 ft would be required to account for the combination of scour and earthquake loads. These results are summarized in Table 13. Clearly there are an infinite number of load factor combinations that would allow us to

150

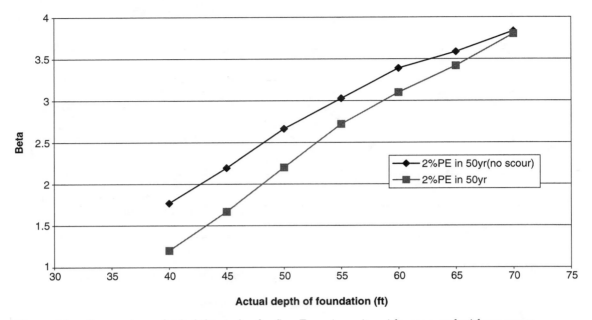

Reliability Index for Soil Failure—Schohaire River Data and San Francisco *EQ* Data

Figure 14.   Comparison of reliability index for San Francisco site with scour and without scour.

TABLE 13   Required foundation depth to satisfy different reliability targets

Target reliability index	Depth for no scour	Depth for scour + *EQ*	Current *EQ* design requirement
1.5	37 ft	43 ft	44 ft
2	43 ft	48 ft	44 ft
2.5	48 ft	53 ft	44 ft
3	55 ft	58 ft	44 ft
3.5	63 ft	65 ft	44 ft

design the foundation to achieve the target reliability levels For example, to achieve the required foundation depth, one option would be to use a load factor on earthquake loads equal to 1.0 and to increment the foundation depth by a factor of 17.34-ft for design scour depth. For example, by taking the required 44 ft depth for earthquake alone with 0.52 times the design scour depth, a total foundation depth of 53 ft is achieved, leading to a reliability index of 2.5. The different load factors that may be used on scour to achieve different reliability targets are listed in the last column of Table 13. If a target reliability index of 2.0 is required, then a load factor on scour of 0.23 will be sufficient.

The results summarized in Table 13 are for illustration only as the data from the other sites and rivers are still being summarized and analyzed. However, it is noted that the differences from the different sites and rivers are expected to be minimal, as demonstrated in the previous sections, in which

it was observed that the reliability indexes calculated using the models described in this report were similar for the different site locations and rivers considered.

REFERENCES FOR APPENDIX C

American Association of State Highway and Transportation Officials (1994). *AASHTO LRFD Bridge Design Specifications,* Washington, DC.

Applied Technology Council (1981). *Seismic Design Guidelines for Highway Bridges,* ATC-6, Redwood City, CA.

Bea, R.G. (1983). "Characterization of the Reliability of Offshore Piles Subjected to Axial Loadings," *Proceedings of the ASCE Structures Congress,* October 1983, TX.

Becker, D.E (1996). "18th Canadian Geotechnical Colloquium: Limit States Design for Foundations, Part II: Development for the National Building Code of Canada," *Canadian Geotechnical Journal,* Vol. 33; pp. 984–1007.

Briaud, J.L., Ting, F.C.K., Chen, H.C, Gudavalli, R., Perugu, S., and Wei, G. (1999). "SRICOS: Prediction of Scour Rate in Cohesive Soils at Bridge Piers," *ASCE Journal of Geotechnical and Environmental Engineering*, Vol. 125, No. 4 (April 1999); pp. 237–246.

Chopra, A.K., and Goel, R.K. (2000). "Building Period Formulas for Estimating Seismic Displacements," *EERI Earthquake Spectra,* Vol. 16, No. 2.

Ellingwood B., Galambos, T.V., MacGregor, J.G., and Comell, C.A. (1980). *Development of a Probability Based Load Criterion for American National Standard A58,* National Bureau of Standards, Washington, DC.

Frankel, A., Harmsen, S., Mueller, C., Barnhard, T., Leyendeker, E.V., Perkins, D., Hanson, S., Dickrnan, N., and Hopper, M. (1997). "USGS National Seismic Hazard Maps: Uniform Hazard Spectra, De-aggregation, and Uncertainty," *Proceedings of FHWA/ NCEER Workshop on the National Representation of Seismic Ground Motion for New and Existing Highway Facilities,* NCEER Technical Report 97-0010, SUNY Buffalo, NY; pp. 39–73.

Haviland, R. (1976). "A Study of Uncertainties in the Fundamental Translational Periods and Damping Values for Real Buildings," MIT reports, Cambridge, MA.

Hwang, H.H.M., Ushiba, H., and Shinozuka, M. (1988). "Reliability Analysis of Code-Designed Structures under Natural Hazards," Report to MCEER, SUNY Buffalo, NY.

Hydraulic Engineering Center (1986). "Accuracy of Computed Water Surface Profiles," U.S. Army Corps of Engineers, Davis, CA.

Johnson, P.A. (1995). "Comparison of Pier Scour Equations Using Field Data," *ASCE Journal of Hydraulic Engineering,* Vol. 121, No. 8; pp. 626–629.

Landers, M.N., and Mueller, D.S. (1996). "Channel Scour at Bridges in the United States," FHWA-RD-95-l84, Federal Highway Administration, Turner-Fairbank Highway Research Center, McLean, VA.

Liu, W.D., Ghosn, M., Moses, F., and Neuenhoffer, A. (2001). *NCHRP Report 458: Redundancy in Highway Bridge Substructures,* Transportation Research Board of the National Academies, Washington, DC.

Liu, D., Neuenhoffer, A., Chen, X., and Imbsen, R. (1998). "Draft Report on Derivation of Inelastic Design Spectrum," Report to NCEER, SUNY Buffalo, NY.

Miranda, E. (1997). "Strength Reduction Factors in Performance-Based Design," EERC-CUREe Symposium in Honor of Vitelmo V. Bertero, Berkeley, CA.

National Earthquake Hazards Reduction Program (1997). *Recommended Provisions for Seismic Regulations for New Buildings and Other Structures,* Federal Emergency Management Agency, FEMA 302, Building Safety Council, Washington, DC.

Newmark, N.M., and Hall, W.J. (1973). "Seismic Design Criteria for Nuclear Reactor Facilities," *Building Practices for Disaster Mitigation, Report No. 46,* National Bureau of Standards, U.S. Department of Commerce; pp. 209–236.

Nowak, A.S. (1999). *NCHRP Report 368: Calibration of LRFD Bridge Design Code,* Transportation Research Board of the National Academies, Washington, DC.

Poulos, H.G., and Davis, E.H. (1980). *Pile Foundation Analysis and Design,* Krieger Publishing Co., FL.

Priestley, M.J.N., and Park, R. (1987). "Strength and Ductility of Concrete Bridge Columns Under Seismic Loading," *ACI Structural Engineering Journal,* January–February.

Richardson, E.V., and Davis, S.R. (1995). *Evaluating Scour at Bridges,* 3rd edition. Report No. FHWA-IP- 90-0 17, Hydraulic Engineering Circular No. 18, Federal Highway Administration, Washington, DC.

Seed, H.B., Ugas, C., and Lysmer, J. (1976). "Site-Dependent Spectator Earthquake-Resistant Design," *Bulletin of the Seismological Society of America,* Vol. 66, No. 1 (February 1976); pp. 221–243.

Shirole, A.M., and Holt, R.C. (1991). "Planning for a Comprehensive Bridge Safety Assurance Program," *Transportation Research Record 1290,* Transportation Research Board of the National Academies, Washington, DC; pp. 39–50.

Stewart, J.P., Seed, R.B., and Fenves, G.L. (1999). "Seismic Soil Structure Interaction in Buildings, II: Empirical Findings." *ASCE Journal of Geotechnical and Geoenvironmental Engineering,* Vol. 125, No. 1.

Takada, T., Ghosn, M., and Shinozuka, M. (1989). "Response Modification Factors for Buildings and Bridges," *Proceedings from the 5th International Conference on Structural Safety and Reliability,* ICOSSAR 1989, San Francisco; pp. 415–422.

Zahn, F.A., Park, R., and Priestly, M.J.N. (1986). "Design of Reinforced Concrete Bridge Columns for Strength and Ductility," Report 86- 7, University of Canterbury, Christchurch, New Zealand.

APPENDIX H

SEISMIC RISK ANALYSIS OF A MULTISPAN BRIDGE

NCHRP Project 12-49 has recently proposed a set of recommended load resistance factor design (LRFD) guidelines for the seismic design of highway bridges, published as *NCHRP Report 472: Comprehensive Specifications for the Seismic Design of Bridges* (Applied Technology Council [ATC] and the Multidisciplinary Center for Earthquake Engineering Research [MCEER], 2002). The proposed LRFD guidelines have adopted and modified many of the features outlined by the existing National Earthquake Hazards Reduction Program specifications (NEHRP, 1997), which were originally intended for buildings, to the design and safety evaluation of bridge structures. Thus, for the life safety limit state, the proposed LRFD seismic guidelines use the NEHRP 2,500-year return period earthquake hazard maps along with the response spectra proposed by NEHRP, but remove a $^2/_3$ reduction factor that is associated with the NEHRP spectral accelerations. The $^2/_3$ factor had been included by NEHRP to essentially reduce the 2,500-year acceleration spectrum to an equivalent 500-year spectrum for the U.S. West Coast region.

Another major change proposed by NCHRP Project 12-49 consists of a new set of response modifications factors that more realistically model the nonlinear behavior of bridge components. Specifically, NCHRP Project 12-49 has proposed higher response modification factors for multicolumn bents, which would reduce the implicit levels of safety in the design process and may offset the use of the more conservative acceleration spectrum. In addition, NCHRP Project 12-49 proposed a performance-based design approach whereby different seismic design and analysis procedures (SDAPs) are specified, depending on the seismic hazard level and whether the bridge is expected to perform adequately for a life safety or operational performance objective (ATC and MCEER, 2002).

The analyses performed in the previous appendixes were executed for bridge piers having foundations formed by pile extensions by modeling the bridge pier as an equivalent single degree of freedom (SDOF) system after determining the effective point of fixity. In this appendix, the analysis of an example bridge founded on a pile system, which had been previously prepared by the NCHRP Project 12-49 team, is reviewed and is used as the basis for the seismic risk analysis. This particular analysis is based on a more advanced multidegree of freedom (MDOF) structural analysis. The objectives of this appendix are (1) to compare the results of the risk analysis obtained using simple SDOF systems with those that using an MDOF spectral analysis and (2) to compare the safety levels of bridge bents founded on multiple piles with those of bridge bents founded on pile extensions.

1. DESCRIPTION OF EXAMPLE BRIDGE

To achieve the objectives of this appendix, an example bridge that was used as part of NCHRP Project 12-49 to illustrate the recommended analysis and design methods is selected for review. Specifically, the bridge selected for this analysis is the 500-ft bridge described in the unpublished Design Example No. 8 developed under NCHRP Project 12-49. The original site was taken by the NCHRP Project 12-49 team to be in the Puget Sound region of Washington state. However, in this appendix, the analysis is repeated for the seismic zone corresponding to Seattle. The change in site location is effected to take advantage of the availability of the Seattle seismic intensity data. The United States Geological Survey (USGS) earthquake hazard maps indicate that, for the Seattle site, the short-period acceleration (0.2 sec) is $S_s = 1.61$ g and the 1.0-sec acceleration is $S_1 = 0.560$ g for the most credible earthquake (MCE) corresponding to the earthquake with a 2,500-year return period.

The 500-ft bridge is formed by five continuous spans of 100 ft each. The four monolithic bents are formed by two columns, each integrally connected to the cap of the combined concrete piles with steel casings that form the foundation system. The two-column bents are also monolithically connected to the box girder superstructure through a crossbeam. Expansion bearings form the connections between the superstructure and the substructure at the two end stub-type abutments. Figures 1.a, 1.b, and 1.c—which are adopted from the NCHRP Project 12-49 example—provide a description of the bridge.

The NCHRP Project 12-49 team chose to use the commercial structural analysis program SAP2000 to perform a multimode spectral analysis of the example bridge. The structural model for each bent is represented by Figure 2, which is adapted from the unpublished NCHRP Project 12-49 report, in which the effect of the foundation is modeled by a series of translation and rotational springs. The same model used in the NCHRP project report is used in this appendix, which focuses on examining the safety of the bridge columns when the bridge is subjected to earthquakes having intensities similar to those expected in Seattle. A separate analysis is executed for each of the transverse and longitudinal directions because the uncertainties associated with the direction of the earthquake and the use of different possible combination rules are beyond the scope of this study.

2. ANALYSIS OF EXAMPLE BRIDGE

The bridge example described in the previous section was analyzed using the same input data for the structural model

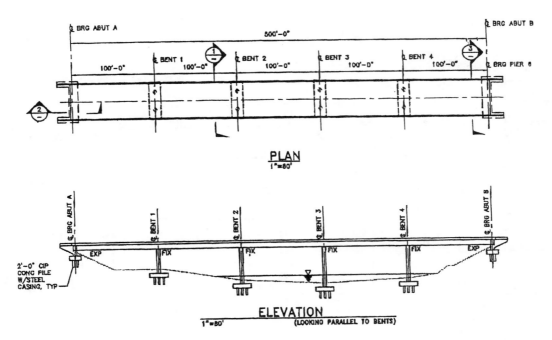

Figure 1.a. Plan view of example bridge.

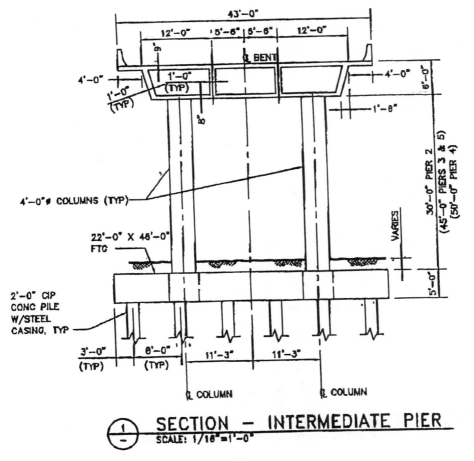

Figure 1.b. Cross section of example bridge.

154

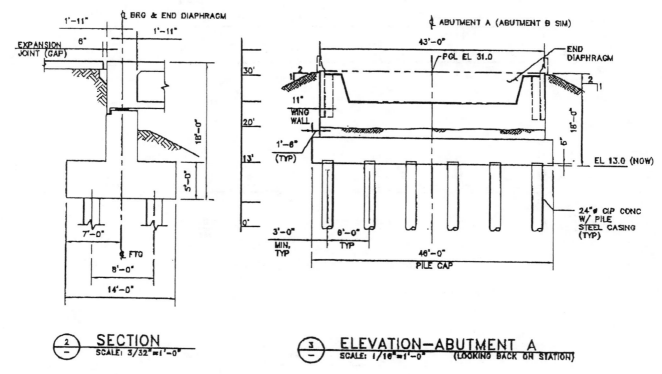

EXPANSION
JOINT (GAP)

1'-11"
6"
1'-11"

₵ BRG & END DIAPHRAGM

30'

20'

13'

0'

18'-0"

5'-0"

₵ FTG

7'-0"

8'-0"

14'-0"

₵ ABUTMENT A (ABUTMENT B SIM)

43'-0"

PGL EL 31.0

END
DIAPHRAGM

2
1

11"

WING
WALL

1'-6"
(TYP)

2
1

18'-0"

6"

EL 13.0 (NOW)

24"ø CIP CONC
W/ PILE
STEEL CASING
(TYP)

3'-0"
MIN,
TYP

6'-0"
TYP

46'-0"
PILE CAP

② SECTION
SCALE: 3/32"=1'-0"

③ ELEVATION—ABUTMENT A
SCALE: 1/16"=1'-0"   (LOOKING BACK ON STATION)

Figure 1.c.   Profile of bridge example abutment.

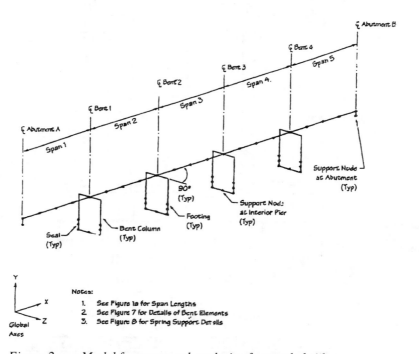

Notes:

1.   See Figure 1a for Span Lengths
2.   See Figure 7 for Details of Bent Elements
3.   See Figure 8 for Spring Support Details

Global
Axes

Figure 2.a.   Model for structural analysis of example bridge.

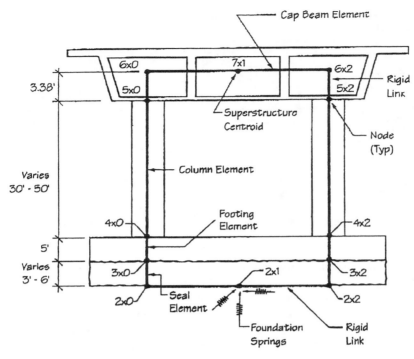

Cap Beam Element

6x0   7x1   6x2
5x0   5x2   Rigid Link

3.38'

Superstructure Centroid

Node (Typ)

Varies 30' - 50'

Column Element

Footing Element

4x0   4x2

5'

Varies 3' - 6'

3x0   2x1   3x2

2x0   Seal Element   2x2

Foundation Springs   Rigid Link

Notes:
1.) Looking Ahead on Line
2.) "x" Represents Pier Number (Pier 2, 3, or 5)

Figure 2.b.   Structural model for column bent with foundation springs.

that was previously used by the NCHRP Project 12-49 team, but using the earthquake response spectrum corresponding to the MCE of the Seattle site. The MCE, as explained by NEHRP (1997) and ATC and MCEER (2002), corresponds to the earthquake with a 2,500-year return period. On the other hand, the AASHTO LRFD specifications specify that bridges should be designed for a 75-year design life. Hence, the difference between the 2,500-year design earthquake return period and the 75-year design life provides an implicit level of safety that must be taken into consideration while performing a seismic risk analysis or when calculating the reliability index for the bridge under consideration. Similarly, the use of response modification factors that are lower than those observed during experimental investigations of bridge members provide additional safety factors. Such implicit safety factors compensate for the use of a load factor equal to $\gamma_{EQ} = 1.0$, which is stipulated by the AASHTO LRFD seismic specifications and other seismic design codes.

The SAP2000 structural model is presented in Figures 2.a and 2.b. The results of the SAP2000 analysis using the MCE for the Seattle site are summarized in Table 1, which shows the maximum moments produced in the bridge columns assuming that the EQ is applied independently in the longitudinal or the transverse directions of the bridge. Using a response modification factor of $R_m = 6.0$, as is done in the NCHRP Project 12-49 example, indicates that the bridge col-

umns should be designed to have moment capacities exactly equal to $M_{xdesign} = 1,606$ kip-ft (= 9638.5 kip-ft/6) for bending about the x axis and $M_{zdesign} = 2,800$ kip-ft (= 16,800 kip-ft/6) for bending about the z axis.

Following traditional practice, the design process outlined above and described in detail by ATC and MCEER (2002) treats all the variables that control the seismic safety of the bridge as deterministic variables. However, it is well known that a large number of uncertainties including modeling and inherent uncertainties are associated with the seismic analysis, design, and bridge construction processes, as well as the estimation of future loads and the identification of material properties. These uncertainties imply that bridges designed and constructed following current procedures are associated with a certain level of risk to failure within their intended 75-year design life. A model to estimate this level of risk as measured by the reliability index, β, is presented in the main body of this report (*NCHRP Report 489*). The model considered that the dominant uncertainties controlling the seismic risk of bridge systems may be included into five random variables that are related to the following:

1. Expected maximum earthquake intensity for the bridge site within the bridge's design life;
2. Natural period of the bridge system;

156

TABLE 1 Summary of SAP2000 results for forces and moments in bridge columns

	Axial Force (kips)	Moment X (kip-ft)	Shear Z (kip)	Moment Z (kip-ft)	Shear X (kip)
EQ Longitudinal					
Bent 1 column	39	0	0	16,800	1,119
Bent 2 column	56	0	0	7,661	340
Bent 3 column	11	0	0	6,239	249
Bent 4 column	33	0	0	7,688	342
EQ Transverse					
Bent 1 column	723	7,544	499	1,089	73
Bent 2 column	778	8,028	355	825	37
Bent 3 column	899	9,639	383	295	12
Bent 4 column	790	8,675	383	785	35

3. Spectral accelerations for the site taking into consideration the soil properties;
4. Nonlinear behavior of the bridge system; and
5. Modeling uncertainties associated with current methods of analysis.

Information available in the literature about these random variables are discussed in detail in the main body of this report and are summarized in the first six rows of Table 2 of this appendix. The distribution of the earthquake accelerations for different sites are also shown in Figure 3, which is based on data provided by USGS and Frankel et al. (1997).

As mentioned above, the structural and reliability analyses used in this report are based on an SDOF model of a bridge

bent. The use of a multimodal analysis for this bridge example implies that the analysis is based on several vibration modes. Therefore, the consideration of a single random variable, namely t', to represent the uncertainties associated with determining the natural period of the bridge is no longer possible. An alternative approach consists of recognizing that the period of the system is a function of the stiffness of the structural members and those of the foundation. The stiffness of each member is controlled by the product of the modulus of elasticity and the moment of inertia. If one assumes that the moment of inertia is deterministic, then the stiffness of the structural member is controlled by one random variable, namely the modulus of elasticity, E. The nominal modulus of elasticity used in design practice is the value obtained for

TABLE 2 Summary of input values for seismic reliability analysis

Variable		Bias	Coefficient of Variation	Distribution Type	Reference
Earthquake modeling, λ_{eq}		1.0	20%	Normal	Ellingwood et al. (1980)
Spectral modeling, C'		1.0	Varies per site (15% to 40%)	Normal	Frankel et al. (1997)
Acceleration A	San Francisco	1.83% g (yearly mean)	333%	from Figure 3	USGS website
	Seattle	0.89% g (yearly mean)	415%		
	Memphis	0.17% g (yearly mean)	1707%		
	New York	0.066% g (yearly mean)	2121%		
	St. Paul	0.005% g (yearly mean)	3960%		
Period, t'		0.90	20%	Normal	Chopra and Goel (2000)
Weight, W		1.05	5%	Normal	Ellingwood et al. (1980)
Response modification, R_m		7.5 (mean value)	34%	Normal	Priestly and Park (1987); Liu et. al (1998)
Modulus of elasticity, E		1.25	40%	Normal	Deduced from Chopra and Goel (2000)
Foundation spring stiffness, K_s		1.0	17%	Normal	Stewart et al. (1999)

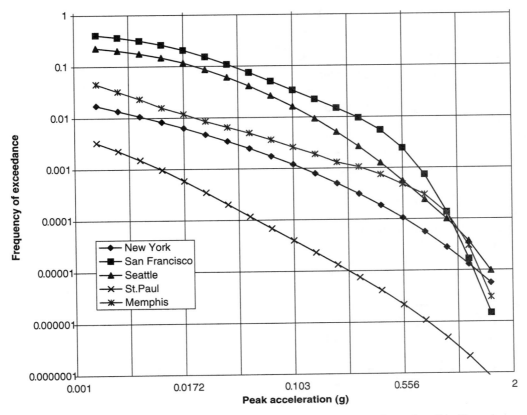

Figure 3. Annual probability of exceedance curves for peak ground acceleration [based on USGS website and Frankel et al. (1997)].

concrete at 28 days. However, some researchers have assumed that the actual long-term modulus for concrete members is on the order of 1.25 times that of the nominal value. For this reason, in the calculations performed in this example, the mean value of E is assumed to be 1.25 times the nominal value. This bias will also result in a bias approximately equal to 0.90 for the natural period of an SDOF system, which corresponds to the period bias used in this report. Using similar logic, it is proposed to use a coefficient of variation (COV) of $V_E = 40\%$ for the modulus of elasticity during the reliability analysis of this bridge example. In fact, $V_E = 40\%$ would result in a COV of 20% for the period of an SDOF system, which is the same value used during the analyses performed in other appendixes of this report. To account for the uncertainties associated with determining the spring stiffness values of the foundation system, the data collected by Stewart et al. (1999) are used. These data indicate that models used to estimate foundation spring coefficients are associated with a bias equal to 1.0 and a COV on the order of 17%.

This report used a closed-form expression to model the SDOF response of bridge piers to earthquake accelerations. Such a closed-form solution lent itself to the use of full-fledged Monte Carlo simulations to study the risk to failure and calculate the reliability index for a variety of bridge configurations. The fact that the response of this bridge example

is not based on an explicit formulation but is obtained implicitly through the use of a commercial analysis program—which requires the interference of the analyst to prepare the input data and interpret the results—precludes the use of the Monte Carlo simulation that requires an extremely large number of analyses and requires the use of more approximate techniques. In particular, the Response Surface Method (RSM) has been used by several researchers to incorporate structural analysis results into structural reliability computations. RSM is based on artificially constructing an equation to model the response of the structure using a polynomial fit to the results obtained from a limited number of discrete numerical analyses. Melchers (1999) observes that when the approximating surface fits the point responses reasonably well in the area around the most likely failure point, RSM provides a good estimate of the reliability index, β. Based on these observations and the difficulties associated with using a Monte Carlo simulation, RSM is used in this report to obtain estimates of the reliability index for the bridge example described above.

The first step in RSM consists of using the structural analysis program SAP2000 at various values of the random variables and of studying the response of the bridge for these input parameters. Melchers (1999) and others (e.g., Ghosn et al., 1994) have indicated that the best points to use during these analyses should be selected such that the points lie close to the most likely failure point. However, since the failure point is not known a priori, an iterative process must be used. In this case, the analysis is first performed at points lying close to the mean values (mean +/− 1 standard deviation) and additional refinements are introduced following the estimation of the location of the most likely failure point. Because the earthquake intensity is associated with an extremely high standard deviation, the SAP2000 analysis is performed at the mean 75-year earthquake acceleration and at points located at +/−0.1 times the standard deviation. The results of the SAP2000 analyses are presented in Table 3. Note that the response of interest to this example is the maximum bending moment in a bridge column. Two cases are considered: (1) the earthquake is applied in the longitudinal direction of the bridge and (2) the earthquake is applied in the transverse direction of the bridge. As mentioned earlier, the two cases are treated separately because the consideration of the earthquake orientation and the directional combination rules are beyond the scope of this study.

The results shown in Table 3 demonstrate that the foundation stiffness has negligible effect on the maximum moment response in the bridge columns. When the earthquake is applied in the transverse direction, no change is observed in the maximum bending moment of the columns. If the earthquake is applied in the longitudinal direction, a change of only 1.7% is observed when the change in the foundation stiffness is equal to 34% (2 times the standard deviations around the mean value). Changes of +/−1 standard deviation in the weight of the structure result in a change of less than 5% in the maximum bending moment. On the other hand, changes between 40% and 50% in the bending moments are observed when the spectral accelerations and the modulus of elasticity are changed by +/−1 standard deviation each. A change in the expected 75-year earthquake peak ground acceleration (PGA) by +/−0.10 standard deviation results in a change of about 13% in the columns' maximum bending moment. When the information summarized in Table 3 is used

in a first order Taylor series expansion, the response surface may be represented by an equation of the following form:

$$M_{\text{analysis transverse}} = -6,409 + 12,523\,EQ + 0.002879\,E \\ + 3,790\,C' + 9,735\,b_W + 0\,b_{Ks} \quad (1)$$

and

$$M_{\text{analysis long}} = -11,765 + 23,322\,EQ + 0.00568\,E \\ + 6,900\,C' + 17,090\,b_W + 0\,b_{Ks} \quad (1')$$

where

$M_{\text{analysis transverse}}$ and $M_{\text{analysis long}}$ = the maximum moment response due to earthquakes in the transverse and longitudinal directions, respectively;

EQ = the earthquake PGA;

E = the modulus of elasticity;

C' = the modeling variable of the spectral acceleration; and

b_W and b_{Ks} = factors that multiply the weights, W, and the foundation stiffness, K_s, respectively.

The normalized factors, b_W and b_{Ks}, are used because each bridge element may have a different weight, and each foundation stiffness my be different. The normalized factors b_W and b_{Ks} in Equations 1 and 1' express how the final moment changes when all the individual weights and stiffnesses are varied by the same factor. When the mean values of the weights are used, b_w is set at 1.0. Similarly, when the mean values of K_s are used, b_{Ks} is set at 1.0. The model assumes that all the element weights are correlated such that a change of the weight of one element by a certain factor leads to the multiplication of all the other elements' weights by the same factor. A similar assumption is made for the foundation stiffnesses.

The results shown in Table 1 indicate that the design procedure proposed in NCHRP Project 12-49 [ATC and MCEER,

TABLE 3 Results of parametric analysis close to mean values

| | *EQ* Transverse | *EQ* Longitudinal |
	Moment about x (kip-ft)	Moment about z (kip-ft)
Base case = analysis at mean values	3,879	6,900
75-yr $EQ + 0.1\,\sigma_{EQ}$	4,054	7,251
75-yr $EQ - 0.1\,\sigma_{EQ}$	3,560	6,331
$C' + \sigma_{C'}$	4,849	8,626
$C' - \sigma_{C'}$	2,954	5,176
$W + \sigma_W$	3,970	7,057
$W - \sigma_W$	3,786	6,734
$E + \sigma_E$	4,595	8,289
$E - \sigma_E$	3,006	5,153
$K_s + \sigma_{ks}$	3,879	6,841
$K_s - \sigma_{ks}$	3,879	6,960

2002) would dictate that the columns be designed to have maximum capacities $M_{\text{cap transverse}} = M_{\text{design transverse}} = 1{,}606$ kip-ft ($9{,}638.5$ kip-ft/R_m) where the $9{,}638.5$ kip-ft is the moment obtained by analyzing the structure under the effect of the 2,500-year earthquake and the nominal values of the material properties and spectral accelerations, and $R_{m \text{ nominal}} = 6.0$ is the nominal response modification factor specified by NCHRP Project 12-49 for multicolumn bents. For the longitudinal direction, the required moment capacity is $M_{\text{cap long}} = M_{\text{design long}} = 2{,}800$ kip-ft.

Failure occurs when the applied moment is higher than or equal to the design moment. The applied moment obtained from the SAP2000 analysis does not account for the nonlinear response that is normally represented using a response modification factor, R_m. In addition, to account for the modeling uncertainties, Ellingwood et al. (1980) have suggested that a modeling factor λ_{EQ} must also be included in the reliability analysis. This would lead to a failure function, Z, that can be represented by the equation

$$Z = M_{\text{cap}} - \lambda_{EQ} M_{\text{analysis}} / R_m \qquad (2)$$

where M_{cap} is a random variable that, according to Nowak (1999), follows a lognormal distribution with a bias equal to 1.14 and a COV of 13%. The modeling factors λ_{EQ} and R_m are also random variables having the properties described in Table 2 above, and M_{analysis} is a function of the random variables EQ, C', E, W, and K_s (or, more precisely, b_W and b_{KS}). In the first iteration, the function that gives M_{analysis} is described in Equation 1 when the bridge is subjected to earthquakes in the transverse direction and in Equation 1' when the bridge is subjected to earthquakes in the longitudinal direction. Monte Carlo simulations using Equations 1 or 1' with Equation 2 produced reliability index values of $\beta = 1.76$ for the transverse direction and $\beta = 1.73$ for the longitudinal direction. Through the review of the various cases that fall within the

TABLE 4 Results of parametric analysis close to most likely failure point

$R_M = 6$	EQ Longitudinal Moment about z (kip-ft)	EQ Transverse Moment about x (kip-ft)
Base case = most likely failure point	18,851	10,722
$EQ^* + 0.1\,\sigma_{EQ}$	19,352	10,986
$EQ^* - 0.1\,\sigma_{EQ}$	18,351	10,444
$C'^* + 0.2\,\sigma_{C'}$	19,714	11,216
$C'^* - 0.2\,\sigma_{C'}$	17,987	10,228
$W^* + 0.2\,\sigma_W$	18,948	10,772
$W^* - 0.2\,\sigma_W$	18,752	10,658
$E^* + 0.2\,\sigma_E$	19,566	11,097
$E^* - 0.2\,\sigma_E$	18,116	10,311
$K_s^* + 0.2\,\sigma_{ks}$	18,814	10,722
$K_s^* - 0.2\,\sigma_{ks}$	18,887	10,722

failure region, the Monte Carlo simulation also indicates that the most likely failure point occurs when $EQ^* = 0.56g$, $E^* = 1.36E_n$ ($b_W^* = 1.05$) or $W^* = W_n$, $b_{C'}^* = 1.07$ and when $b_{Ks}^* = 1.0$ or $K_s^* = K_{sn}$ for the earthquake in the transverse direction. The maximum likely failure point occurs when $EQ^* = 0.60g$, $E^* = 1.39E_n$, $W^* = 1.05W_n$, $C'^* = 1.08$, and $K_s^* = K_{sn}$ when the earthquake is applied in the longitudinal direction. These most likely failure points are then used to perform a second iteration of the parametric analysis with results described in Table 4. These will in turn lead to new sets of response functions similar but with slightly different coefficients than those in Equations 1 and 1'.

The use of the results presented in Table 4 along with Equation 2 within a second and third reliability analysis leads to an updated reliability index of $\beta = 1.76$ for the transverse direction and 1.74 for the longitudinal direction. The analysis is then repeated for various values of M_{cap} and is illustrated in Figure 4.

Reliability Index versus Moment Capacity

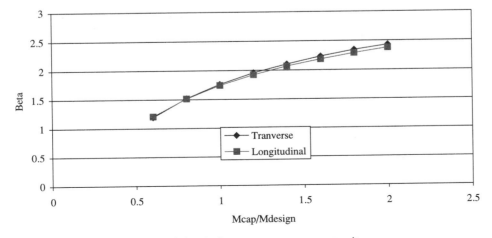

Figure 4. Variation of reliability index versus moment capacity.

The results of Figure 4 show that the reliability index values for the longitudinal and transverse directions produce a reliability index on the order of 1.75 if the proposed design guidelines of NCHRP Project 12-49 were to be adopted ($M_{cap}/M_{design} = 1.0$) (ATC and MCEER, 2002). The reliability index would increase to close to 2.35 when the moment capacity of the columns is selected to have capacities equal to twice the recommended design capacities ($M_{cap}/M_{design} = 2.0$). This increase in the reliability index could then be achieved by, for example, requiring the use of a response modification factor during the design process of $R_m = 3.0$ rather than the $R_m = 6.0$ that has been recommended by NCHRP Project 12-49. It is to be noted that because the analysis accounted for the nonlinear behavior of the members, the calculations executed herein correspond to the reliability of the bridge system. The reader is reminded that reliability index on the order of $\beta = 3.5$ has been used as the basis for the design of bridge members under gravity loads. The presence of system redundancy would increase the system reliability index under gravity to $\beta > 4.0$. The lower reliability index values for earthquake loads is justified in the engineering community on the basis of an informal cost-benefit analysis whereby expert chief bridge engineers contend that there is general satisfaction with current design procedures given the excessive economic costs that would be associated with increasing the current reliability levels.

The results in Figure 4 show that the reliability index values are lower than those calculated in the main body of this report (*NCHRP Report 489*). This observed difference is due to the fact that the analysis performed herein is for the bridge columns while the analysis in the main body of the report was for the extension pile in the bridge foundation. The difference is primarily caused by the use of a lower response modification factor of $R_m = 1.5$ for the extension pile as opposed to the $R_m = 6.0$ used herein for bridge columns supported on multi-pile foundations.

To study the effect of the nominal value of the response modification factor R_m used for design on the reliability index, the reliability analysis of this five-span bridge is repeated for different values of the response modification factor and compared with the three-span bridge example analyzed in the main body of the report. The results for the case in which the earthquake is applied in the transverse direction are compared in Figure 5.

The results shown in Figure 5 demonstrate how the reliability index increases with the use of higher response modification factors and how the results obtained from this example based on an MDOF SAP2000 analysis of a multispan bridge founded on multi-pile foundation systems approach the results obtained in the body of the report for bridges on extension piles analyzed as SDOF systems. It is clear from the results that the major difference is due to the use of different response modification factors for different bridge members.

3. CONCLUSIONS

This appendix executed a reliability analysis for the seismic response of a five-span bridge with columns supported on multi-pile foundations and compared the results with those obtained in other parts of this report for bridges analyzed as SDOF systems. The results illustrate the following points:

1. The simplified analysis using an SDOF system yields results consistent with those obtained from the more advanced multimodal structural analysis.
2. RSM yields reasonable approximations to the reliability index for the seismic risk analysis of simple bridge configurations.
3. The use of a response modification factor of $R_m = 6.0$ for bridge columns associated with the 2,500-year NEHRP spectrum produces system reliability index values on the order of 1.75. This is much lower than the member

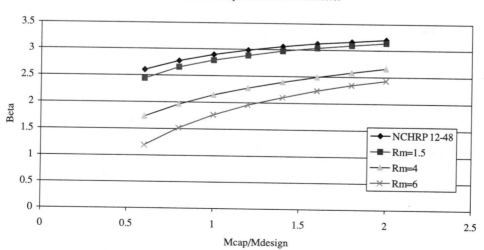

Figure 5. Effect of response modification factor on reliability index.

reliability equal to 3.5 used as the basis for calibrating the LRFD code for members under gravity load.

4. Large differences in the reliability indexes are observed for the different components of bridge subsystems subjected to seismic events. For example, the foundation systems produce much higher reliability index values than do the bridge columns. This difference has been intentionally built into the bridge seismic design process by the code writers (i.e., the NCHRP Project 12-49 research team) in order to account for the consequences of the failure of different members. However, the difference is rather large, producing a beta that varies from a value of about $\beta = 3.0$ when a response modification factor equal to $R_m = 1.0$ is used to about $\beta = 1.75$ when a response modification factor of $R_m = 6.0$ is used during the design process. It is noted that the use of a response modification factor of $R_m = 1.0$ implies a design based on the elastic behavior of members while an $R_m = 6.0$ accounts for the plastic behavior of bridge columns.

5. Although not common in the realm of seismic bridge engineering, the use of the reliability index as a measure of bridge safety for all types of extreme events will provide a consistent measure that will help in calibrating load combination factors as executed during the course of this study (see Chapter 3 of this report).

REFERENCES FOR APPENDIX H

Applied Technology Council and the Multidisciplinary Center for Earthquake Engineering Research (2002). *NCHRP Report 472: Comprehensive Specification for the Seismic Design of Bridges,* Transportation Research Board of the National Academies, Washington, DC.

Chopra, A.K., and Goel, R.K. (2000). "Building Period Formulas for Estimating Seismic Displacements," *EERI Earthquake Spectra,* Vol. 16, No. 2.

Ellingwood B., Galambos, T.V., MacGregor, J.G., and Comell, C.A. (1980). *Development of a Probability Based Load Criterion for American National Standard A58,* National Bureau of Standards, Washington, DC.

Frankel, A., Harmsen, S., Mueller, C., Barnhard, T., Leyendeker, E.V., Perkins, D., Hanson, S., Dickrnan, N., and Hopper, M. (1997). "USGS National Seismic Hazard Maps: Uniform Hazard Spectra, De-aggregation, and Uncertainty," *Proceedings of FHWA/ NCEER Workshop on the National Representation of Seismic Ground Motion for New and Existing Highway Facilities,* NCEER Technical Report 97-0010, SUNY Buffalo, NY; pp. 39–73.

Ghosn, M., Moses, F., and Khedekar, N. (1994). "Response Function and System Reliability of Bridges,"*Probabilistic Structural Mechanics: Advances in Structural Reliability Methods*; Springer-Verlag; pp. 220–236.

Liu, D., Neuenhoffer, A., Chen, X., and Imbsen, R. (1998). "Draft Report on Derivation of Inelastic Design Spectrum," Report to NCEER, SUNY Buffalo, NY.

Melchers, R.E. (1999). *Structural Reliability: Analysis and Prediction,* John Wiley & Sons, New York, NY.

National Earthquake Hazards Reduction Program (1997). *Recommended Provisions for Seismic Regulations for New Buildings and Other Structures,* Federal Emergency Management Agency, FEMA 302, Building Safety Council, Washington, DC.

Nowak, A.S. (1999). *NCHRP Report 368: Calibration of LRFD Bridge Design Code,* Transportation Research Board of the National Academies, Washington, DC.

Priestley, M.J.N., and Park, R. (1987). "Strength and Ductility of Concrete Bridge Columns Under Seismic Loading," *ACI Structural Engineering Journal,* January–February.

Stewart, J.P., Seed, R.B., and Fenves, G.L. (1999). "Seismic Soil Structure Interaction in Buildings, II: Empirical Findings." *ASCE Journal of Geotechnical and Geoenvironmental Engineering,* Vol. 125, No. 1.

APPENDIX I

ANALYSIS OF SCOUR DATA AND MODIFIED RELIABILITY MODEL FOR SCOUR

1. INTRODUCTION

The AASHTO LRFD specifications (1998) state that "a majority of bridges that have failed in the United States and elsewhere have failed due to scour." This is confirmed by Shirole and Holt (1991), who observed that over the last 30 years, more than 1,000 of the 600,000 U.S. bridges have failed and that 60% of these failures are due to scour while earthquakes accounted for only 2%. Of course, there are many more bridges that are posted or otherwise taken out of service due to their inadequate strengths (e.g., due to deterioration, low rating, fatigue damage, etc.); nevertheless, scour is considered a critical cause of failure because its occurrence often leads to catastrophic collapses. For this reason, developing methods for the design and maintenance of bridge foundations subjected to risk from scour as well as developing tools to detect the presence of scour are currently considered top priorities for agencies concerned with the safety of bridges.

Scour is typically assessed using the methodologies presented in FHWA's Hydraulic Engineering Circular No. 18 (Richardson and Davis, 1995), known as "HEC-18." Although the HEC-18 model requires designing bridges to sustain foundation scour associated with the water flow depth and velocity corresponding to the 100-year or overtopping flood event, the remaining variables in the HEC-18 equations are considered to be deterministic. In other words, once the 100-year water depth and flow velocity have been estimated, the HEC-18 methodology, like most typical design procedures, uses a deterministic design approach with safety factors that implicitly account for uncertainties in the models, the parameters, or the hydraulic and hydrologic variables. The HEC-18 design equation for scour depth around bridge piers is given as follows:

$$\frac{y_{s\,design}}{y_1} = 2.0 K_1 K_2 K_3 K_4 \left(\frac{D}{y_1}\right)^{0.65} F_0^{0.43} \qquad (1)$$

where

$y_{s\,design}$ = the nominal or design scour depth;
y_1 = the upstream flow depth;
D = the effective pier width;
F_0 = Froude number; and
K_1, K_2, K_3 and K_4 are correction factors for the pier shape, angle of attack, bed forms, and sediment gradation, respectively.

The flow depth, y_1, and the flow velocity used to calculate the Froude number are those corresponding to the 100-year or overtopping flood. The safety factors may be included in the

bias factor of 2.0, the correction factors, K_i, and/or the usage of 100-year return period for y_1, which is longer than the 75-year design period stipulated for the design of bridges under the LRFD specifications.

If Equation 1 is developed to provide a safe (or nominal) value for scour depth, the actual depth of scour, $y_{s\,scour}$, may be obtained from the following:

$$\frac{y_{s\,actual}}{y_1} = 2.0\lambda K_1 K_2 K_3 K_4 \left(\frac{D}{y_1}\right)^{0.65} F_0^{0.43} \qquad (2)$$

where λ is a modeling factor which has an average value equal to 0.55 as reported by Johnson in Appendix B of this report.

If a bridge foundation is designed to satisfy Equation 1 and the actual depth of scour is obtained from Equation 2, the margin of safety, Z, can be defined as follows:

$$Z = y_{s\,design} - y_{s\,actual}. \qquad (3)$$

Failure would occur when $y_{s\,actual}$ is greater or equal to $y_{s\,design}$ or when the margin of safety, Z, is less than or equal to zero. Once the designer chooses the parameters of Equation 1 following the HEC-18 guidelines, $y_{s\,design}$ will be exactly known. On the other hand, because of the randomness in the scour process, the determination of the actual scour depth $y_{s\,actual}$ is associated with large uncertainties, and $y_{s\,actual}$ must be considered to be a random variable. The sources of randomness are primarily due to the following.

- The modeling uncertainties associated with the HEC-18 equation, which does not provide very accurate estimates of scour depths even if all the input data are precisely known.
- The uncertainties associated with predicting the input data. These, as identified by Johnson (see Appendix B), are primarily due to the following:
  - The determination of the maximum 75-year discharge rate.
  - The determination of the flow depth and flow velocity corresponding to the 75-year discharge rate. Because for known channel profiles the flow and flow velocity are obtained from the discharge data based on Manning's roughness coefficient, the uncertainties are assumed to be primarily due to the determination of the appropriate Manning coefficient to use.
  - The random variation in river bed conditions and its influence on scour depth as represented by the factor K_3 in the above equations.

The prediction of the 75-year probability of failure, P_f, is defined as the probability that $Z \leq 0$ and can be estimated based on the statistical data describing the randomness of each of the variables that affect Equation 3, as described above and as listed in Table 1.

Although safety factors (or biases) may have been introduced into the empirically derived HEC-18 models, there is always a finite chance that the actual scour depth is larger than the design scour depth. This is due to the uncertainties associated with predicting the maximum flow depth and velocity that the river will be exposed to within the bridge's design period and the uncertainties associated with determining the correction factors, K_i, as well as the modeling errors that are introduced by the HEC-18 equations. The object of this appendix is to provide an explicit accounting for the uncertainties and the implied risks of the current scour design methodology in order to verify the compatibility between the safety of bridges subjected to scour and other loads and extreme events. To address issues related to structural risk, recent developments in bridge design procedures have used a reliability-based methodology to calibrate load and resistance safety factors. The goal of the calibration is to provide a consistent level of risk for the range of pertinent bridge geometries and loading conditions. The nominal measure of risk that is used during the calibration process is the reliability index, β, that is mathematically related to the probability of failure, P_f, by

$$\beta = \Phi^{-1}(P_f) \qquad (4)$$

where Φ^{-1} is the inverse function of the cumulative normal distribution.

The reliability index, β, would give an exact measure of risk if all the random variables that affect the safety of the bridge component under consideration are properly identified along with their probability distribution types and pertinent statistics. Because of the difficulty of calculating P_f exactly, in most practical situations, the reliability index is calculated using approximate methods such as first or second order reliability method (FORM or SORM) algorithms or Monte Carlo

simulations. Also, because many of the random variables that control the probability of failure are difficult to identify, categorize, and estimate and also because the statistical database for civil engineering applications is usually very limited, the reliability calculations would provide only a nominal measure of risk rather than an actuarial value. These limitations produce what is generally known as "modeling uncertainties."

To reduce the effects of the above-mentioned modeling uncertainties, the calibration of new design codes is often executed to match the reliability index obtained from existing "satisfactory" designs rather than satisfying a predetermined specific value for the probability of failure. This approach has worked reasonably well when calibrating load and resistance factors for bridges having typical geometries and materials having a long history of satisfactory performance under reasonably predictable loading conditions and when the reliability index, β, has been used as the primary criterion for developing new design codes. Analyses performed during the course of this study have demonstrated that the reliability index is highly controlled by the statistical properties of the modeling variable λ. For this reason, this appendix will focus on studying the modeling uncertainties associated with the scour design procedures and the influence of these uncertainties on the estimates of the reliability index. This analysis is based on the data provided by Landers and Mueller (1996) in their study supported by the FHWA.

2. SCOUR MODELING UNCERTAINTIES

For the reliability analysis of bridge piers subjected to scour, a Monte Carlo simulation can be used to generate random samples for each of the random variables that control the safety margin Z of Equation 3 based on their specified probability distributions. The estimation of the statistics including mean values and coefficients of variation (COVs) of the pertinent variables as well as the probability distribution types depend on each bridge's pier geometry, channel configuration, and channel bed properties. Evidently, several difficulties arise when attempting to estimate the properties of the

TABLE 1   Input data for reliability analysis for scour alone

Variable	Mean Value	Coefficient of Variation	Distribution Type	Reference
Q, discharge rate	From Table 2.8	From Table 2.8	Lognormal	Based on USGS website
λ_Q, modeling variable for Q	1.0	5%	Normal	Based on USGS website
λ_{sc}, modeling variable	0.55	52%	Normal	Johnson (1995)
n, Manning roughness	0.025	28%	Lognormal	Hydraulic Engineering Center (1986)
K_3, bed condition factor	1.1	5%	Normal	Johnson (1995)

variables that control scour at specific bridge sites, particularly when the bridge has not yet been constructed and actual measurements on the maximum scour depth are not available, even over limited periods of time. Even when such measurements may be available, the projection of the expected scour depth over the full design life of the bridge is associated with large levels of uncertainty. Added to these difficulties is the realization that the HEC-18 equations were primarily developed based on a limited number of small-scale laboratory experiments under "ideal" flow conditions and uniform sand bed materials that may not resemble actual field conditions. In fact, several investigations have shown large discrepancies between observed scour depths and those predicted from the HEC-18 equations. Because of all the limitations of the HEC-18 model, a procedure that takes into consideration the modeling uncertainties in an explicit manner while studying the safety of bridges subjected to scour is initiated in this appendix.

In assessing the safety of bridge piers subjected to scour, four broad sources of uncertainty are relevant: (1) inherent variability, (2) estimation error, (3) model imperfection, and (4) human error. Inherent variability, often called "randomness," may exist in the characteristic of the bridge structure or in the environment to which the structure is exposed. Inherent variability is intrinsic to nature and is beyond the control of the engineer or code writers. The uncertainties due to estimation error and model imperfection are extrinsic and, to some extent, reducible. For example, a reliability analyst may choose to obtain additional information to improve the accuracy of estimation. The uncertainty due to human error may also be reduced by implementing rigorous quality-control measures during the collection, interpretation, and analysis of scour data. Such improvements, however, usually entail an investment in time and money that the analyst may not be willing to undertake. In addition, this effort requires the capability of collecting such data at the site of interest. During the bridge design process when the bridge is still in the planning stages, the collection of site-specific data is not possible. Hence, it is usual to rely on design models developed based on data from other sites and for different bridge configurations and to project these available data to the particular site and bridge conditions. Hence, improving the quality of the data collection effort is not a viable option during the design stages, although this can be achieved at later stages when checking the safety of existing bridges. This study focuses on developing load factors for the design of new bridges under extreme events. Thus, the study must rely on available data collected from a variety of sites under different conditions to obtain estimates of the modeling uncertainties that describe the relationship between the expected scour depths and those predicted from the HEC-18 equations. In bridge engineering practice, pier scour depth is estimated using the HEC-18 model with Equation 1.

Field observations have shown that even when all the pertinent parameters of HEC-18 are known, large discrepancies still exist between the measured scour depths and those of the HEC-18 equations. These differences are attributed to the uncertainties in the HEC-18 model. To account for these modeling uncertainties, Equation 1 can be modified by inserting a model correction factor, λ, which is defined as follows:

$$\lambda = \frac{y_{s\,actual}}{y_{s\,design}} \tag{5}$$

so that the actual depth of scour, $y_{s\,actual}$, may be obtained from Equation 2. Notice that λ as defined in Equation 5 provides an inverse measure of safety such that a lower value of λ implies a higher level of safety. Also, since λ is a random variable, the higher its COV is, the higher the probability is that the actual depth of scour will exceed the design value and, hence, the lower the safety of the system is.

The mean value of λ is defined as $\bar{\lambda}$ and its standard deviation as σ_λ. These can be estimated from observations of scour depths at bridge piers and by comparing the observed scour depths with those predicted using the HEC-18 equation. For example, a 1996 study supported by FHWA has collected an extensive set of data on observed scour depths and compared these with different prediction models including the HEC-18 equations. The data published in the report by Landers and Mueller (1996) provides $y_{s\,actual}$ and sufficient information to calculate $y_{s\,design}$ for different sites, channel bed materials, and types of bridge piers. Table 2 provides a summary of the average values of λ and its COV for several different categories and groupings of the data.

A total number of 374 local scour measurements at different bridge piers were assembled from the report of Landers and Mueller (1996). Of these measurements, 240 were collected in channels with live-bed conditions and the rest were for piers set in clear-water conditions. Live-bed conditions occur when the water channel carries soil particles that are deposited on the channel floor at low flow speeds. These deposits would refill the scour hole such that under live-bed conditions, the scour process becomes cyclical. From Table 2, it is observed that the COV for the data collected in live-bed channels is lower than that obtained from all the data. Out of the 240 measurements taken in live-bed channels, 126 were for piers with rounded noses, 52 of the piers had square noses, 30 piers were cylindrical, and 32 had sharp noses. On the other hand, 191 of the bridges had single-pier bents and 49 had pier groups. The bridge piers were also classified based on three types of foundations: (1) piles, (2) poured, and (3) unknown. The soil type is classified as noncohesive soil or unknown soil. Only 18 data points satisfied the conditions of cylindrical single poured piers in live-bed channels having noncohesive soils. When all the data are analyzed simultaneously, the COV is 64.6%, indicating a very high level of modeling uncertainties. Large differences in the mean values

TABLE 2 Summary of mean and COV of λ based on data of Landers and Mueller (1996)

Flow and channel material type	Pier shape	Mean	Standard deviation	COV	Number of observations	Standard deviation of mean
All cases		0.412	0.266	0.646	374	0.0138
Channels with live-bed conditions only		0.429	0.247	0.576	240	0.0159
All channel bed material	Rounded	0.400	0.231	0.577	126	0.0206
	Sharp	0.523	0.292	0.558	32	0.0516
	Cylinder	0.383	0.204	0.532	30	0.0372
	Square	0.432	0.246	0.570	52	0.0341
Noncohesive soils	All shapes	0.417	0.237	0.569	195	0.0170
Unknown soil type	All shapes	0.479	0.283	0.593	45	0.0422
Single piers	All shapes	0.405	0.223	0.550	191	0.0161
Pier groups	All shapes	0.535	0.310	0.580	49	0.0443
Pile foundation	All shapes	0.421	0.256	0.607	158	0.0204
Poured foundation	All shapes	0.419	0.185	0.442	67	0.0226
Unknown foundation	All shapes	0.547	0.361	0.660	15	0.0932
Noncohesive soils, poured	Rounded	0.405	0.181	0.446	48	0.0261
Noncohesive soils, poured	Cylinder	0.355	0.132	0.371	18	0.0311

and/or COVs are observed for different pier shapes, foundation types, and so forth. Lower COVs are observed when the data are categorized based on pier shape and streambed conditions. As an example, the mean value of λ for the 18 cases that have cylindrical poured piers in noncohesive soils was found to be 0.36 with a COV of 37.1%. The piers with rounded noses produced a mean value of λ equal to 0.41 with a COV equal to 44.6%. The lowest value of COV is observed for the case of cylindrical piers in poured foundations set in live-bed channels with noncohesive soils. Presumably, this lower COV of 37.1% reflects the fact that the HEC-18 model was primarily developed based on laboratory simulations under these same conditions. Although the lowest in Table 2, a COV of 37.1% still indicates a high spread in the observed data away from the mean value. The bias of 0.36 for this case implies a conservative design equation with an implicit safety factor on the order of 2.82 (1/0.36).

The point estimates of the mean and the standard deviations of λ given in Table 2 are obtained from a limited set of data. To convey information on the degree of accuracy of these estimates, one should recognize that λ is a random variable with (unknown) mean μ_λ and with (unknown) standard deviation σ_λ. The sample means that values of $\bar{\lambda}$ given in Table 2 are estimates of μ_λ for various data sets. The precision of these estimates can be assessed using the fact that a sample mean $\bar{\lambda}$ is a random variable with mean value equal to μ_λ and a standard deviation of $\sigma_{\bar{\lambda}} = \sqrt{(\sigma_\lambda)^2/n}$ where n is the sample size. It is noted that for large sample sizes, $\bar{\lambda}$ will approach a normal distribution, and the confidence intervals for $\bar{\lambda}$ may be obtained using the normal probability tables. The data of Landers and Mueller (1996) also provide estimates of σ_λ (column 4 standard deviation) from which $\sigma_{\bar{\lambda}}$ is calculated as given in the last column of Table 2. Notice how, in many instances, as the classification of the data is narrowed to spe-

cific categories of pier shapes and water channel conditions, the standard deviation (σ_λ) decreases; however, since the number of samples used to calculate $\bar{\lambda}$ decreases also, the standard deviation of $\bar{\lambda}$ (i.e., $\sigma_{\bar{\lambda}}$) increases, thus expressing a lower level of precision in the estimation of $\bar{\lambda}$.

The data from all 374 cases are plotted in a histogram, as is shown in Figure 1. The figure also shows the histograms that would be obtained if the data were assumed to follow a normal distribution or a lognormal distribution. The plots seem to indicate that the modeling variable λ may be reasonably well modeled by a lognormal distribution. Further verification of this observation for different subsets of the data is undertaken below.

The data from the live-bed channels consisting of 240 data points are also plotted using a lognormal probability scale, as is shown in Figure 2. The plot further verifies that λ for this set of data may be reasonably well represented by a lognormal distribution. Other probability plots for different sets of data are provided in Figures 3, 4, and 5. In particular, Figure 3 plots the modeling factor λ for rounded poured piers set in live-bed channels with noncohesive soils on a normal probability curve. Figure 4 plots the same data on a lognormal probability scale. Figure 5 gives a plot on a lognormal scale for the data assembled for cylindrical poured piers in live bed-channels with noncohesive soils. It is noted that Figures 3 and 4 show that the distribution of λ for the rounded piers approaches a normal distribution near the upper part of the curve (i.e., for high values of λ) while for low values of λ, the probability distribution more closely approaches that of a lognormal distribution. On the other hand, a lognormal distribution seems to be appropriate for the cylindrical piers, as shown in Figure 5.

The chi-squared goodness-of-fit tests for the data sets plotted in Figures 2 through 5 were executed as summarized in Tables 3 through 6. The results confirm the observations made

Comparison of actual data with assumed distribution

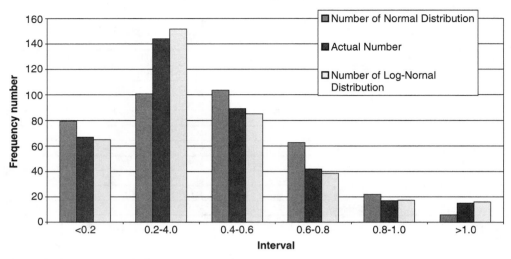

Figure 1. Histogram for distribution of λ.

from the probability plots. In particular, the chi-squared test for the complete data as depicted in Table 3 shows that the data are not inconsistent with the assumption that the underlying distribution is lognormal with a significance level of 5%. In fact, the sum of the squared difference between the observed frequencies and the corresponding lognormal frequencies divided by the lognormal frequency produced a sum equal to 1.02, which is lower than the value of the appropriate χ_f^2 distribution at the cumulative probability of 95%, which is given in probability tables as $C_{1-\alpha, f} = 7.81$ where $\alpha = 5\%$ and $f = 3$. On the other hand, the assumption of a normal distribu-

tion produces a sum of the squared difference equal to 45.86, which is much higher than $C_{1-\alpha, f} = 7.81$. Table 4 shows the same results for the data set consisting of 240 samples collected at sites with live-bed conditions. Table 5 gives the results for the scour-modeling ratio, λ, for the 48 rounded poured piers set in noncohesive soils in live-bed channels. Table 6 gives the same analysis for the cylindrical piers. It is noted that for all the cases considered, the lognormal distribution is acceptable within the 5% confidence level. However, Table 5 shows that for the rounded piers, the normal distribution would also be valid.

Log-normal probability paper

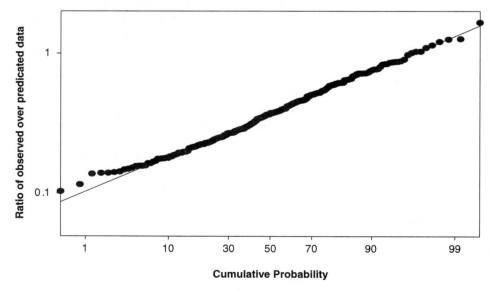

Figure 2. Plot of data from live-bed conditions on lognormal probability paper.

Normal probability paper

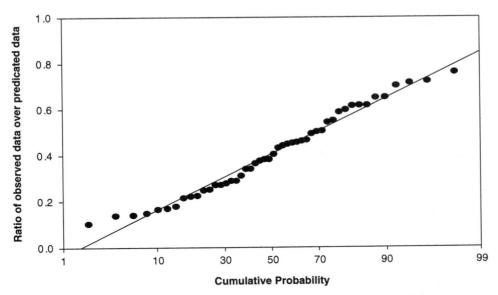

Figure 3. Plot of data for rounded piers under live bed on normal probability paper.

Currently used reliability analysis procedures account for modeling uncertainties by including λ in the failure function and treating it as a random variable in a similar way as all the other physical variables that affect the safety of a structural system. However, the use of a single modeling variable λ for all scour cases assumes that the expected ratio between the observed scour and the scour predicted from the HEC-18 equations is constant for all levels of scour and is independent of any of the parameters that control scour at a given site. This is not necessarily the case. A plot of λ versus the logarithm of the observed scour depth is shown in Figure 6. The plot shows how the modeling variable varies with the observed scour depth for the 48 samples corresponding to scour in noncohesive soil around rounded poured piers. The

Log-normal probability paper

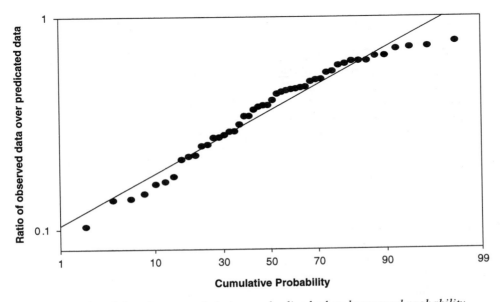

Figure 4. Plot of data from rounded piers under live bed on lognormal probability paper.

168

Log-normal probability paper

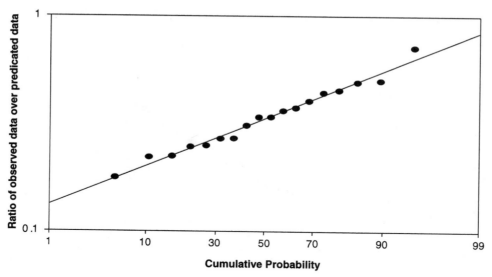

Figure 5. *Plot of data for cylindrical piers under live-bed conditions on lognormal probability paper.*

TABLE 3 Chi-squared goodness-of-fit test for λ using all scour data

Frequency Range	N_i Observed	e_i, Normal Distribution	e_i, Lognormal Distribution	$(N_i-e_i)^2/e_i$ for Normal	$(N_i-e_i)^2/e_i$ for Lognormal
< 0.2	67	79.5	65	1.97	0.06
0.2–0.4	144	100.6	151.7	18.72	0.39
0.4–0.6	89	103.5	85.1	2.03	0.18
0.6–0.8	42	62.7	38.5	6.83	0.32
0.8–1.0	17	22	17.3	1.14	0.01
> 1.0	15	5.7	16	15.17	0.06
Sum	*374*			*45.86*	*1.02*
$C_{1-\alpha,f}$				*7.81*	*7.81*

TABLE 4 Chi-squared goodness-of-fit test for λ using scour data from live-bed channels

Frequency Range	N_i Observed	e_i, Normal Distribution	e_i, Lognormal Distribution	$(N_i-e_i)^2/e_i$ for Normal	$(N_i-e_i)^2/e_i$ for Lognormal
< 0.2	36	30.7	29.8	0.91	1.29
0.2–0.4	100	86.5	109.2	2.10	0.77
0.4–0.6	55	89.1	62.5	13.02	0.89
0.6–0.8	29	30.3	24.1	0.05	0.98
0.8–1.0	11	3.35	8.8	17.50	0.53
> 1.0	9	0.12	5.6	660.64	2.06
Sum	*240*			*694.22*	*6.51*
$C_{1-\alpha,f}$				*7.81*	*7.81*

plot of Figure 6 clearly shows skewness in λ. Other plots and a regression analysis indicate that λ increases asymptotically as the observed scour depth increases. The relationship between λ and the scour depth may be represented by an equation of the following form:

$$\lambda = 0.24 \, y_s^{0.49}. \tag{6}$$

The standard error for the above equation is found to be $\varepsilon = 0.464$ and the multiple $R^2 = 0.42$. Alternatively, a multivariable regression analysis of λ as a function of the pier width, D, flow depth, y_1, and flow velocity, V, shows that λ can be represented in an equation of the form

$$\lambda = 0.184 + 0.0081D + 0.0044 \, y_1 + 0.014 \, V. \tag{7}$$

TABLE 5 Chi-squared goodness-of-fit test for λ using scour data of rounded piers in live-bed channels

Frequency Range	N_i Observed	e_i, Normal Distribution	e_i, Lognormal Distribution	$(N_i-e_i)^2/e_i$ for Normal	$(N_i-e_i)^2/e_i$ for Lognormal
< 0.2	7	6.14	5.96	0.12	0.18
0.2–0.3	10	7.30	11.25	1.00	0.14
0.3–0.4	7	10.00	10.59	0.90	1.21
0.4–0.5	9	10.16	7.60	0.13	0.26
0.5–0.6	6	7.65	4.89	0.36	0.25
0.6–0.7	5	4.28	3.01	0.12	1.32
> 0.7	4	2.47	4.71	0.95	0.11
Sum	48			3.58	3.47
$C_{1-\alpha,f}$				9.49	9.49

TABLE 6 Chi-squared goodness-of-fit test for λ using scour data of cylindrical piers in live-bed channels

Frequency Range	N_i Observed	e_i, Normal Distribution	e_i, Lognormal Distribution	$(N_i-e_i)^2/e_i$ for Normal	$(N_i-e_i)^2/e_i$ for Lognormal
< 0.2	1	2.15	1.31	0.62	0.07
0.2–0.3	6	3.93	5.52	1.09	0.04
0.3–0.4	5	5.31	5.66	0.02	0.08
0.4–0.5	4	4.16	3.22	0.01	0.19
0.5–0.6	1	1.88	1.41	0.41	0.12
0.6–0.7	0	0.49	0.55	0.49	0.55
> 0.7	1	0.08	0.33	10.54	1.36
Sum	18			13.18	2.42
$C_{1-\alpha,f}$				9.49	9.49

for 48 samples

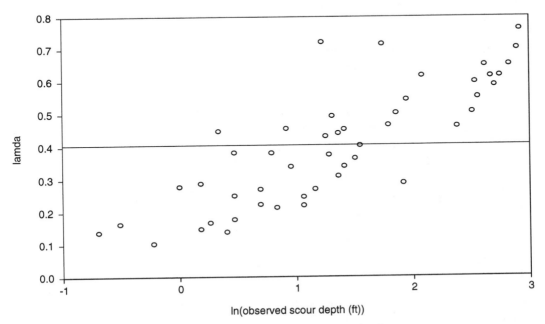

Figure 6. Plot of modeling variable, λ, versus observed scour depth.

170

The above equation is executed for the 48 samples of rounded piers in noncohesive soils under live-load conditions. In fact, using an analysis of variance, it was found that the effect of the pier shape on the average value of λ was not statistically significant given the high observed standard deviations.

To account for the skewness in λ that is caused by the use of the HEC-18 equations, an alternative method of analysis is suggested as described in Section 3 of this appendix.

3. ALTERNATIVE RELIABILITY MODEL FOR SCOUR

Appendix C of this report presented the results of a reliability analysis of bridge piers subjected to scour. The model used in Appendix C was based on data that were previously available in the literature and assumed that the modeling variable λ is uniform for all cases. The existence of the extensive database published by Landers and Mueller (1996) provides a unique opportunity to come up with an alternative empirical model for predicting scour depths based on a multivariable regression analysis. Several alternative regression models have been analyzed, including a regression effected on the whole data set consisting of 374 samples, on the 48 samples for the rounded piles in noncohesive soils, and on the combination of rounded and cylindrical piers.

Based on the parameters tabulated by Landers and Mueller (1996) and after several trials to identify the parameters that control the scour depth, the regression equation for the total 374 sample size was selected to be of the following form:

$$\ln (\text{scour}) \sim \text{shape} + \text{bed-load transport} \\ + \ln (\text{pier width}) + \ln (\text{flow depth}) \\ + \ln (\text{flow velocity}) \tag{8}$$

where ln is the log function. The regression coefficients obtained using the S+ computer package are tabulated as

shown in Table 7; the residual standard error was found to be 0.56, and the term $R^2 = 0.53$.

The regression analysis of the 48 samples of rounded piers in noncohesive soils was performed using a model of the following form:

$$\ln.(\text{scour}) \sim \ln (\text{pier width}) + \ln (\text{flow depth}) \\ + \ln (\text{flow velocity}) \tag{9}$$

In this case, the regression coefficients are listed in Table 8. The residual standard error was found to be 0.41 on 44 degrees of freedom and the term R^2 is 0.83.

The analysis of the residuals for the 48 sample points as plotted in Figures 7 through 9 shows that the points are evenly spread around the value of 0.0 and that they can be reasonably well represented by a normal distribution.

The use of the regression results in the reliability analysis may be effected by replacing the failure function of Equation 3 by the equation

$$Z = y_{\text{design}} - y_{\text{regression}} \tag{10}$$

where for rounded piers in noncohesive soils $y_{\text{regression}}$ is calculated from

$$\ln y_{\text{regression}} = -2.08 + 0.63 \ln D + 0.48 \ln y_1 \\ + 0.61 \ln V + \varepsilon \tag{11}$$

where

D = the pier diameter,
y_1 = the flow depth,
V = the flow velocity, and
ε = the residual error.

TABLE 7   Coefficients of regression analysis for all 374 samples of data

Coefficients	Value	Standard Error	t Value	Pr (> |t|)
(Intercept)	−0.60	0.10	−6.10	0.000
Shape 1	−0.19	0.11	−1.68	0.094
Shape 2	−0.10	0.04	−2.47	0.014
Shape 3	−0.02	0.03	−0.67	0.500
Shape 4	0.02	0.02	0.96	0.338
Bed.load	0.16	0.03	4.60	0.000
ln.width.b	0.66	0.07	9.58	0.000
ln.flow.depth	0.20	0.04	5.04	0.000
ln.flow.vel	0.12	0.05	2.15	0.032

TABLE 8   Coefficients of regression analysis for 48 samples of rounded pier data

Coefficients	Value	Standard Error	t Value	Pr (> |t|)
(Intercept)	−2.08	0.30	−6.98	0.0000
ln.width.b	0.63	0.14	4.62	0.0000
ln.flow.depth	0.48	0.13	4.10	0.0002
ln.flow.vel	0.61	0.13	4.78	0.0000

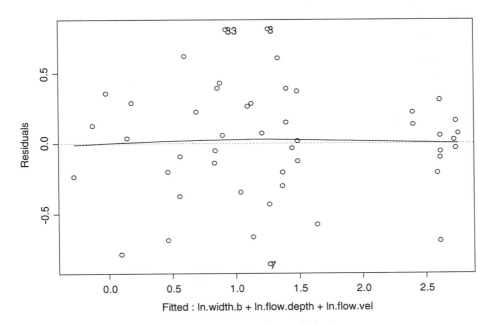

Figure 7. Plot of residuals for 48 samples of rounded piers.

As explained above, *D* is a deterministic variable since the pier diameter can be accurately known even before the actual construction of the pier; y_1 and *V* are random variables that depend on Manning's roughness coefficient and the 75-year maximum discharge rate, which are random variables having the properties listed in Table 8. Based on the analysis of the residuals effected in this section of the report, ε may be considered to follow a normal distribution with mean equal to zero and a standard deviation equal to 0.41.

Using Equation 10 in a Monte Carlo simulation as described in Section 2 of Appendix C of this report for the Schohaire

River, data would lead to the results shown in Figure 10. The plot shows that if the HEC-18 equation is used for designing the bridge pier, then the critical design scour depth should be $y = 17.3$ ft. The reliability index obtained for such a design is $\beta = 0.47$. The reliability index increases to $\beta = 2.71$ when the scour depth is designed to be twice the value of HEC-18 (i.e., $y_{actual}/y_{HEC-18} = 2.0$ or $y_{actual} = 34.6$ ft).

A slight difference in the results is observed when $y_{regression}$ is calculated based on the regression equation obtained when the full set of 374 samples is used (i.e., from Equation 8). These results are also shown in Figure 10. In this case, the

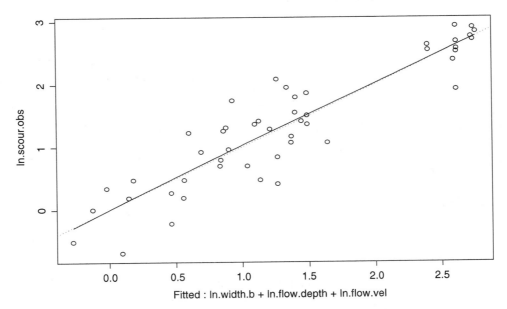

Figure 8. Comparison of regression results to observed scour depth.

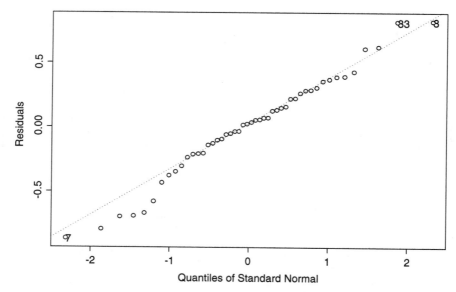

Figure 9. Plot of residual versus standard normal quantiles.

reliability index of β = 0.58 is obtained when the foundation is designed for a critical depth of 17.32 ft as stipulated from the HEC-18 equations. The reliability index is 2.45 when the foundation depth is set at 34.64 ft. The small differences observed between the results obtained from Equation 8 and Equation 9 are due to the larger standard regression error obtained from Equation 8 that uses all 378 sample points.

The reliability analysis is subsequently performed for the five different rivers that were analyzed in Appendix C: the Schohaire, Mohawk, Sandusky, Cuyahoga, and Rocky Rivers. These relatively small rivers are selected because they pro-

duce 100-year flood depths ranging from 8.5 ft to 21 ft, which are appropriate for the bridge configuration under investigation as described in Appendix C. As was done in Appendix C, two cases are considered: (1) the case in which the bents are formed by one column each and (2) the case in which the bents are formed by two columns. The results of the reliability analysis for each case are presented in Figures 11 and 12.

Tables 9 and 10 also provide a summary of the results obtained using this alternative reliability model for the bridge with the one-column bent and for the two-column bent. The results illustrate how the reliability index decreases for the

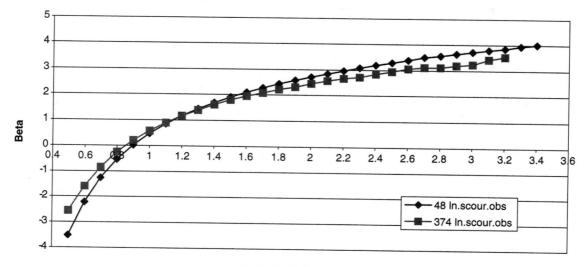

Figure 10. Plot of reliability index versus foundation depth.

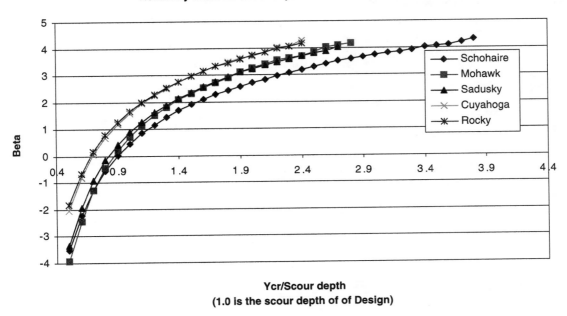

Figure 11. Plot of reliability index for one-column bent for five river discharges.

rivers with higher discharge rates varying from $\beta = 1.66$ for the Rocky River to $\beta = 0.47$ for the Schohaire River. The lower beta for the rivers with higher discharge rates reflects the influence of a lower modeling variable, λ, for high values of scour depths. Such results indicate that the HEC-18 equation becomes less conservative as the expected scour depth increases. The average reliability index for the cases consid-

ered is on the order of 1.08. If one is to group the rivers into three categories—high discharge rates (Schohaire), medium discharge rates (Mohawk and Sandusky), and small discharge rates (Cuyahoga and Rocky)—then the average beta would be close to $\beta = 1.0$. This value is clearly lower than the $\beta = 1.40$ observed in Appendix C when the reliability model employing a uniform modeling variable is used.

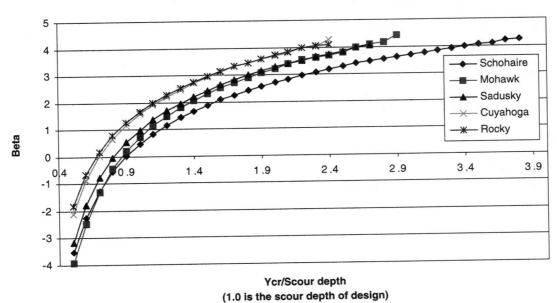

Figure 12. Plot of reliability index for two-column bent for five river discharges.

174

TABLE 9 Summary of reliability analysis for one-column bent for different river discharge data

River	Average Ys75 (max scour depth in 75 yr)	COV of Ys75	Design depth (ft)	Reliability index, β
Schohaire	16.29	25%	17.3	0.48
Mohawk	12.56	21%	14.0	0.73
Sandusky	12.03	22%	14.3	0.90
Cuyahoga	8.88	21%	12.3	1.60
Rocky	8.68	22%	12.3	1.66

TABLE 10 Summary of reliability analysis for two-column bent for different river discharge data

River	Average Ys75 (max scour depth in 75 yr)	COV of Ys75	Design depth (ft)	Reliability index, β
Schohaire	11.60	25%	12.3	0.47
Mohawk	8.95	21%	9.9	0.71
Sandusky	8.57	22%	10.2	0.99
Cuyahoga	6.33	21%	8.7	1.57
Rocky	6.18	22%	8.75	1.65

4. CONCLUSIONS

This appendix presented an analysis of the scour data published by Landers and Mueller (1996). A new reliability model was then proposed based on the results of the analysis. The reliability calculations have shown that the HEC-18 model provides varying levels of reliability depending on the expected scour intensity at a site. In particular, the use of HEC-18 for designing the foundations of bridge piers provides higher safety levels for rivers with relatively small discharge rates (differences in the reliability index on the order of 1.20 are observed for the sites selected in this appendix). In all cases, however, the reliability index observed for bridge foundations designed according to the HEC-18 equations are much lower (on the order of β = 1.0) than those of bridge members designed for gravity loads using the current AASHTO LRFD specifications, which produce a reliability index on the order of 3.5. Adding a load safety factor on the HEC-18 equations will help increase the average reliability index as observed in Section 3.2 of the main body of this report. However, a major review of the HEC-18 equations will be needed in order to reduce the spread in the observed reliability levels between different sites.

The model proposed herein is based on a regression fit for the data of Landers and Mueller (1996), which is limited to sites with observed scour depths less than 20 ft. The model may not be appropriate for use for sites at which the scour depth is expected to exceed this value.

REFERENCES FOR APPENDIX I

American Association of State Highway and Transportation Officials (1998). *AASHTO LRFD Bridge Design Specifications,* 2nd edition, Washington, DC.

Hydraulic Engineering Center (1986). "Accuracy of Computed Water Surface Profiles," U.S. Army Corps of Engineers, Davis, CA.

Johnson, P.A. (1995). "Comparison of Pier Scour Equations Using Field Data," *ASCE Journal of Hydraulic Engineering,* Vol. 121, No. 8; pp. 626–629.

Landers, M.N., and Mueller, D.S. (1996). "Channel Scour at Bridges in the United States," FHWA-RD-95-184, Federal Highway Administration, Turner-Fairbank Highway Research Center, McLean, VA.

Richardson, E.V., and Davis, S.R. (1995). *Evaluating Scour at Bridges,* 3rd edition. Report No. FHWA-IP-90-017, Hydraulic Engineering Circular No. 18, Federal Highway Administration, Washington, DC.

Shirole, A.M., and Holt, R.C. (1991). "Planning for a Comprehensive Bridge Safety Assurance Program," *Transportation Research Record 1290,* Transportation Research Board of the National Academies, Washington, DC; pp. 39–50.